AF232137

ENVIRONMENTAL SYSTEMS AND SOCIETIES

Internal Assessment

Environmental Systems and Societies

Internal Assessment

For the International Baccalaureate Diploma

Zouev Elite Publishing

Published 2021

Printed by Zouev Elite Publishing

ISBN 978-1-9996115-4-5, paperback.

TABLE OF CONTENTS

PART I
THE IB ESS IA GUIDE

1. GENERAL INTRODUCTION

Internal Assessment (IA) is an integral part of the Environmental Systems and Societies (ESS) curriculum contributing an important **25%** towards the final grade. It is designed to enable students to learn and integrate skills and knowledge learned from the course and apply them in a practical manner to answer personal inquisitions while staying with-in the limitations of the curriculum. An IA involves an individual investigation into a certain aspect of the syllabus using skills learnt and presenting it in form of a written report. This investigation needs to be carried and written following certain guidelines provided and must be done exclusively for the purpose of an IA (An extended essay cannot be used as an IA).

The teaching time allocated for an IA is usually **10 hours** this not only reflects the time recommended for interaction between student and teacher but also includes the time spent by the student in performing and writing the IA task. However, most students end up spending more than 10 hours.

The IA task itself can be from any part of the ESS syllabus but must be a student's original work. Help regarding deciding a topic and planning the task can be sought from the teacher. It is teacher's job to advise students about marking guidelines and IB policies before the start an IA. However, it is not a teacher's job to tell a student exactly what to do or write any part of the report.

Before we get into detail of assessment criteria let's discuss some housekeeping rules of an ESS IA:

- All IAs are internally graded by teachers. Some of these IAs (or all IAs for some new schools/teachers) will also be graded externally by IA moderators/examiners. This external moderation enables a quality control for International Baccalaureate Organization (IBO).
- There is a **word limit** (but no page limit). The report must be between 1500 to 2250 words and it is better to mention the word count on the title page. The moderator/teacher is recommended to not read anything beyond the word limit this can drastically affect your score. Details of word count are also discussed in the 'communication' section of assessment criteria.
- Both primary and secondary data sources can be used for an IA. There is no preference. In addition, it is possible to collect data in groups as part of some activity. You also share data with other students but be very careful to make sure rest of the IA is your original work and not inspired by their work.

2. CHOOSING A TOPIC

Choosing a topic can be tricky for certain students. ESS IA requires you to pick any **Environmental Issue (EI)** related to the ESS syllabus that may be global or local in nature. However, it should involve both Human and Science parts of the ESS curriculum; since ESS in an interdisciplinary course. It is highly recommended to choose a topic that interests you and develop a reasonable **Research Question (RQ)** from it. A lot hangs on your choice of EI and the related RQ so spend good time finalising these.

Figure 1: It would be better to pick a global issue that also has a local component (2).

Once you have picked an EI and RQ, start thinking about how you would answer the RQ. A wide range of methodologies can be used in ESS IAs but they must be appropriate to answer the RQ and you need to consider the logistical details into account as well. Pick a topic and RQ that can be answered within the scope of the IA. Although you can use quantitative and qualitative data, and any statistical analysis techniques, make sure you have read the IB guidelines, in particular go through IB's Animal experimentation policy (3). It would be best to avoid working with live animals but if you must read the above-mentioned policy guides and make sure you are not infringing any.

Let's discuss some possible examples of IA topics categories, please do note these are only suggestion and range of accepted topics is extremely wide:

- **Environmental Impact Assessment (EIA):**

  If there a new construction in your area that you have access to you can do an EIA of your own. Getting permission may be your first hurdle but these EIAs are good way studying any new addition to your school/college. You may look at impact of construction (scope), predict future impact on society, and/or analyse possible solution of mitigating impact. The exact data required would depend on the RQ but you may have to gather primary data from the contractor and survey the society around it. You may support this with secondary data of the area or from similar previous projects.

Figure 2: Environmental impact assessment of a local construction site can be an interesting topic to work on (4).

- **Investigate the changes in an area caused by human activity:**

  This include studies of biodiversity and human activities and allows you to investigate a local issue. This category is very popular in students wanting to do field work. You can gather primary data to compare before-after of the same location or two similar location to investigate effect of human activities. You must narrow down your topic to a specific area, effect, and type of human activity you are studying. Example might include studying biodiversity of plants/insect in disturbed and undisturbed areas, effects of ecotourism etc. Be careful how you word your RQ because that will decide what type to data you must collect and not all types of data are equally easy to collect, this may include:

  - plant species abundance
    - density
    - frequency
    - percentage coverage
    - species diversity (calculated using Simpson diversity index)
  - infiltration rate of the soil
  - amount of litter
  - number of people that pass by

- **Investigate Environmental Value Systems (EVS) or Ecological Footprints (EF):**

  A survey-based approach to investigate EVS or EF of a certain group/population is often a popular IA topic among students. Usually students tend to corelate one or more variables with EVS/EF. These are often simple to design. However, a focused question and well-designed survey can improve your score a lot. While doing this category of IA a local approach is preferable and care must be taken to not overestimate relationships between variables.

3. ASSESSMENT CRITERIA AND GRADING:

For the ESS Internal assessment a total of 6 assessment criteria are used for a total of 30 points. All criteria do not carry equal points.

Indentifying the context	Planning	Results, analysis and conclusion	Discussion and evaluation	Applications	Communication	Total
6 (20%)	6 (20%)	6 (20%)	6 (20%)	3 (10%)	3 (10%)	30 (100%)

Each assessment criterion has level descriptors that outline specific grade points range. All grading is done based on these descriptors, therefore these need to be of foremost importance while writing your report. Each individual criterion has 2-3 aspects that an examiner/teacher would consider while grading. Although each of these aspects may be looked at separately the overall grade of the criterion is awarded using a best fit approach. So, a mark is awarded based on overall fit of the work in a specific level although some aspects of the work may fall below or above the mark level. The highest level of mark does not imply a perfect work. An IA can score full marks even with minor imperfections. Moreover, each of these criteria are judged independently of each other; a high score in one does not imply a high score in other.

Let's get into specific details of these level descriptors for each criterion and discuss some tips for maximising score.

Identifying the context:

This criterion assesses the extent to which the student establishes and explores an environmental issue (either local or global) for an investigation and develops this to state a relevant and focused research question.

Achievement level	Descriptor
0	The student's report does not reach a standard described by any of the descriptors given below.
1–2	The student's report: • **states** a research question, but there is a lack of focus • **outlines** an environmental issue (either local or global) that is linked to the research question • **lists** connections between the environmental issue (either local or global) and the research question but there are significant omissions.
3–4	The student's report: • **states** a relevant research question • **outlines** an environmental issue (either local or global) that provides the context to the research question • **describes** connections between the environmental issue (either local or global) and the research question, but there are omissions.
5–6	The student's report: • **states** a relevant, coherent and focused research question • **discusses** a relevant environmental issue (either local or global) that provides the context for the research question • **explains** the connections between the environmental issue (either local or global) and the research question.

There are 3 aspects to this criterion: **Research Question (RQ), Environmental issue (EI), and a connection between the RQ and EI.** The primary focus is on relevance and coherence of these 3 aspects. The IA revolves around a relevant Environmental issue and must be linked appropriately with a very well thought out research question. This section of the IA is probably the most important since this defines the theme and limits of your IA. Make sure to make the RQ as narrow and focused as possible while still being able to establish a link with the EI. The EI itself needs to be discussed in detail and the link with the RQ must be explained with reasoning. Let's look at some important points to consider while writing this section to maximise your score:

- The RQ must be **relatively specific** for full marks. If the investigation involves studying the effect of watering peas with acidified water then the question should be "What is the effect of watering peas with acidified water on numbers of seeds germinated?" rather than "What is the effect of acid rain on seed germination?"

- **Example RQs:** the following RQs could all be appropriate if the student links them to an EI (not just a human to human issue but some environmental link is needed) or they could have no EI, making the RQ inappropriate.
    - ❖ How does artificial fertilizer affect organisms feeding on a soccer field?
    - ❖ To what extent can abortion affect the population of Mexico City in the future?
    - ❖ What are the health impacts of noise pollution in a city?

- Unfocused, but relevant, RQs are often one of the following:
    - ❖ too vague on the variables to be studied.
    - ❖ too broad in the scope of the 2250 words.
    - ❖ too broad for a 10-hour investigation.
    - ❖ e.g In what ways does water pollution affect an area? Is too broad and too vague - what area and what specific pollutants? The topic is relevant to the EI of pollution.

❖ e.g What impact do humans have on air quality? Is too broad and too vague.
- There must be **evidence of some research** by the student for full marks, especially when they are dealing with a well-studied topic. There must be some depth in the context. For example, a study of rabbit populations in Chile (rabbits are an invasive, introduced species) should include some information on the impact of introduced species in general and specifically in Chile (or state that such information is not available).
- Avoid **general statements** that are common knowledge, with no evidence of personal research.
- When the investigation **does not mention a relevant environmental issue**, the maximum mark awarded in this criterion is a 2. For example, "Stress in schools related to workload" is not a relevant EI. A more subtle example might be a practical to study the effect of light colour on photosynthetic rates. This may or may not relate to an environmental issue; if the student does not identify the link to an EI, then it would score a maximum of 2.

Planning:

This criterion assesses the extent to which the student has developed appropriate methods to gather data that is relevant to the research question. This data could be primary or secondary, qualitative or quantitative, and may utilize techniques associated with both experimental or social science methods of inquiry. There is an assessment of safety, environmental and ethical considerations where applicable.

Achievement level	Descriptor
0	The student's report does not reach a standard described by any of the descriptors given below.
1–2	The student's report: • **designs** a method that is inappropriate because it will not allow for the collection of relevant data • **outlines** the choice of sampling strategy but with some errors and omissions • **lists** some risks and ethical considerations where applicable.
3–4	The student's report: • **designs** a repeatable* method appropriate to the research question but the method does not allow for the collection of sufficient relevant data • **describes** the choice of sampling strategy • **outlines** the risk assessment and ethical considerations where applicable.
5–6	The student's report: • **designs** a repeatable* method appropriate to the research question that allows for the collection of sufficient relevant data • **justifies** the choice of sampling strategy used • **describes** the risk assessment and ethical considerations where applicable.

As shown in the descriptor there are 3 aspects to consider: **Method, Sampling Strategy, and Risk/Ethical Considerations.** Following a similar theme as previous criterion the emphasis is on appropriate method of research. This method should link with your RQ and must fully answer that particular question. Although the method used may vary for each IA and RQ the basic parameters must remain the same.

The method used must be repeatable and it must collect sufficient and reliable data to reach an eventual conclusion. The sampling strategy must be justified in detail to support your choice of method used. Finally, there should be consideration of all risks involved in the method used and the ethical implications involved including getting consent from any human subjects involved. Let's look at some important points to consider while writing this section to maximise your score:

- **Variables** are not necessary, but if included they should match the RQ and the method.
- **Repeats** (or comparison of sources) are nearly always necessary to establish reliability. The appropriate number of repeats is dependent on the investigation carried out. Many standard lab-based reports should plan to have at least 5 repeats. Fieldwork comparing areas should have at least 5+ samples from the two areas, this could be a transect with 5 quadrat samples in each area. Survey and secondary data usually require 20-30 data points as a minimum, this can depend on the statistical processing. (In practice, this may not be possible.)
- Sufficient data to permit processing, should be collected. However, the quantity of data collected will vary with the investigation. It is for the teacher/examiner to judge whether the investigation is collecting insufficient amount of data or not.
- Where relevant safety, ethical and environmental issues would not have been immediately obvious to a student without a teacher's prior knowledge and experience students are given a benefit of doubt.
- If there is a **risk/ethical** issue, then stating it is evidence of listing by the student but does not mean that it has been described. It would of more benefit to describe few important risks in detail than to just list a lot of them.
- For **Surveys**: those that indicate the respondents have been selected randomly is a description at best, but not a justification of the sampling strategy. To justify this, you have to reason why this was done.
- For **field research** it is important to justify the choice of area selected in context of the RQ.

Results, analysis and conclusion:

This criterion assesses the extent to which the student has collected, recorded, processed and interpreted the data in ways that are relevant to the research question. The patterns in the data are correctly interpreted to reach a valid conclusion

Achievement level	Descriptor
0	The student's report does not reach a standard described by any of the descriptors given below.
1–2	The student's report: • **constructs** some diagrams, charts or graphs of quantitative and/or qualitative data, but there are significant errors or omissions • **analyses** some of the data but there are significant errors and/or omissions • **states** a conclusion that is not supported by the data.
3–4	The student's report: • **constructs** diagrams, charts or graphs of quantitative and/or qualitative data which are appropriate but there are some omissions. • **analyses** the data correctly but the analysis is incomplete • **interprets** some trends, patterns or relationships in the data so that a conclusion with some validity is deduced.
5–6	The student's report: • **constructs** diagrams, charts or graphs of all relevant quantitative and/or qualitative data appropriately • **analyses** the data correctly and completely so that all relevant patterns are displayed • **interprets** trends, patterns or relationships in the data, so that a valid conclusion to the research question is deduced.

As clear from the descriptor this criterion revolved around **processing, presentation, and interpretation of data** to reach a conclusion. Although the descriptor seems straight forward this is probably a commonly misinterpreted criterion for students. You should always start with presentation of your raw data (collected data) in some form of table. The emphasis in this section is not on the complexity of your data analysis skills but on using the data appropriately to reach a conclusion. Therefore, processes your raw data and show how and why you processed it in a certain way. Then construct all relevant diagrams or graphs from this processed data and not the raw data. You may analyse the data statistically or visually, but it must identify and all relevant pattern that relate to your RQ. Finally, a clear conclusion based on your data must be provided. This conclusion is not an evaluation and is entirely based on your data. Let's look at some important points to consider while writing this section to maximise your score:

- Quantitative data is usually more significant, **qualitative data** is only required where it is relevant to allow the research question to be answered.
- **Raw data** should be included in the main body of the report and not in the appendix.
- If only pie charts are given, in the absence of raw data, for example in surveys or data for footprint calculations, it is impossible to determine if the data have been analysed correctly. e.g Online surveys from platforms like Googlesheets or Surveymonkey produce these sorts of charts. The students would need to have the raw numbers and not just the graphs.
- The data presented is accepted at **face value**, even if it seems unbelievable. e.g seeds producing flowers after two days. There is no deduction of marks based on raw data. It is the further processing and use of this data which carries significance.
- Good achievement will be possible in investigations where the student has averaged the dependent variable data, validated the processed data (standard deviation/best fit line) and constructed appropriate diagrams/charts/graphs to see the nature of the relationship.

- Correctly tabulated data with appropriate titles and in context must be presented so that teachers/examiners can follow the processing and reasoning, and verify the validity of the interpretations and calculations.
- Students can use graphing calculators or statistical programs for the calculations, examples of full **calculations** are not required for full marks, as long as the calculation is specified.
- If the student has performed a **survey** and the only evidence of analysis is the generation of google doc pie charts, or such like, this does not constitute a complete analysis.
- A statement of whether the data answers the RQ should be presented. It should refer back to the RQ.
- The **conclusion** should be based on the data presented from the investigation. if the conclusion is scientifically wrong this should be discussed in the evaluation section of next criterion.

Discussion and evaluation

This criterion assesses the extent to which the student discusses the conclusion in the context of the environmental issue, and carries out an evaluation of the investigation

Achievement level	Descriptor
0	The student's report does not reach a standard described by any of the descriptors given below.
1–2	The student's report: • **describes** how some aspects of the conclusion are related to the environmental issue • **identifies** some strengths and weaknesses and limitations of the method • **suggests** superficial modifications and/or further areas of research.
3–4	The student's report: • **evaluates** the conclusion in the context of the environmental issue but there are omissions • **describes** some strengths, weaknesses and limitations within the method used • **suggests** modifications and further areas of research.
5 – 6	The student's report: • **evaluates** the conclusion in the context of the environmental issue • **discusses** strengths, weaknesses and limitations within the method used • **suggests** modifications addressing one or more significant weaknesses with large effect and further areas of research.

This criterion is usually the least well scored in the ESS IA. This stems from lack of understanding, and differentiation of conclusion in the previous section and evaluation in this section. They may seem similar but in fact these are two entirely different aspects. **Do not merge the conclusion and evaluation into one section.** You are required to evaluate the conclusion from the previous section in the context of Environmental issue and RQ. This is where you discuss the scientific validity of your conclusion and any limitations of your conclusion. You can talk about to what degree is your conclusion valid and what factors may contribute to limiting the conclusion and what it might mean in context of EI and RQ. You can use external sources here. You may use these external resources to refute your conclusion. In addition to evaluating the conclusion you are required to also discuss Strengths, weaknesses and limitations of the method used. Note that this aspect deals with the method and is different from previous aspect which deals with the conclusion which may not have involved comments on method used. Last aspect here asks for suggested modifications to the method that improve any of the weaknesses/limitations mentioned in the previous aspect. This is separate from suggesting a further area of research. Let's look at some important points to consider while writing this section to maximise your score:

- If there are no references to the **original question** or context, this will be treated as if an aspect is totally missing, i.e. automatically drop an entire band
- Obvious weaknesses that are consistent with the method should be identified and addressed. Consideration of both the strengths and limitations, and the quality of the data and processing is needed
- If a student reflects on how his/her conclusions could be more valid and justified by adapting the method to address underlying factors such as range, sample size, the use of an alternative system to study the same phenomenon, etc, this is considered **an evaluation of the methodology not the conclusion.**
- The evaluation of the nature of the experimental errors (random or systematic) or weaknesses should be consistent with the findings and not **overstated**. As part of this process, the student shows an understanding of the source of errors and his/her relative impact on the reliability of the findings. Comments on significance of sources of error must be consistent with the method.
- A student who only addresses practical or **procedural issues,** by simply giving an account of how his/her results could be improved by carrying out the stated procedure better, would not score full marks.
- Errors due to poor manipulative skills or hypothetical events for which there is no evidence given by the candidate fits maximum 4 for this aspect.
- Do not miss out any one element of strengths, weaknesses and limitations.
- Modifications must be related to the weaknesses identified, contribute to the reliability, precision and accuracy of the results
- **Further research** should be qualitatively different, i.e. collecting more samples is not enough whereas undertaking research in different seasons would be.
- The further research suggested should follow on from the research in a meaningful way and go beyond the original method of investigation show how it will enhance understanding of the topic or RQ

Applications:

This criterion assesses the extent to which the student identifies and evaluates one way to apply the outcomes of the investigation in relation to the broader environmental issue that was identified at the start of the project

Achievement level	Descriptor
0	The student's report does not reach a standard described by any of the descriptors given below.
1	The student's report: • **states** one potential application and/or solution to the environmental issue that has been discussed in the context • **describes** some strengths, weaknesses and limitations of this solution.
2	The student's report: • **describes** one potential application and/or solution to the environmental issue that has been discussed in the context, based on the findings of the study, but the justification is weak or missing • **evaluates** some relevant strengths, weaknesses and limitations of this solution
3	The student's report: • **justifies** one potential application and/or solution to the environmental issue that has been discussed in the context, based on the findings of the study • **evaluates** relevant strengths, weaknesses and limitations of this solution.

Yes, there is a separate applications criterion which needs to be addressed separately. Some students tend to either not include any application or assume they have discussed it with some previous section. It is better to have a dedicated section for this criterion since it is not too difficult to score in. As written in the descriptor a potential application or solution to the EI needs to be justified and discussed with reasoning and it must be based on the findings of the study. This does not mean this has to agree with your conclusion but must address the EI. The solution must be directly linked to the EI - when a solution is suggested that does not match the data and the EI a zero is awarded. This discussed application is then evaluated similarly to your conclusion in the previous criterion. Use a few sentences to address the strengths, weaknesses and limitations of the application. Do not skip any. A good application section uses 2-3 sentences discussing the solution and then completely evaluates it.

Communication:

This criterion assesses whether the report has been presented in a way that supports effective communication in terms of structure, coherence and clarity. The focus, process and outcomes of the report are all well presented

Achievement level	Descriptor
0	The student's report does not reach a standard described by any of the descriptors given below.
1	• The investigation has limited structure and organization. • The report makes limited use of appropriate terminology and it is not concise. • The presentation of the report limits the reader's understanding.
2	• The report has structure and organization but this is not sustained throughout the report. • The report either makes use of appropriate terminology or is concise. • The report is mainly logical and coherent, but is difficult to follow in parts.
3	• The report is well structured and well organized. • The report makes consistent use of appropriate terminology and is concise. • The report is logical and coherent.

Communication is one criterion which is applied globally. Although teachers/examiners area advised to use a global approach to all criteria that is not always the actual practice. This particular criterion however deals with the whole report and should be easy to fulfil. A well-structured report that is easy to follow and uses appropriate scientific terminology and conventions of the field, and is within the word limit would score well here. Many students will not be working in their first language and any errors of expression and spelling made by students would not impact on the mark unless it causes ambiguity, or it becomes incomprehensible. The **structure, relevance and conciseness** of the report are important rather than the language used. With that let's look at some important points to consider while writing this section to maximise your score:

- The style of presentation of the methodology is flexible with no prescription in issues of passive versus personal voice or prose versus a stepwise procedure.
- Structure is evidenced by the use of headings and subheadings, a logical structure and use of diagrams and images to enhance interpretation.
- An investigation that is clear will be coherent. It will be easy to understand the thinking process used by the candidate throughout. This may include showing clearly presented raw and processed data and sample calculations (e.g. steps involved in data processing).
- The contents of an **appendix** are not given credit.
- The work should remain closely connected and appropriate to the topic and question under investigation. It should retain this focus throughout, making it easy to follow the development of the ideas from beginning to end.
- A **descriptive** essay, regardless of coherence, structure, etc. would receive a low mark.
- High achievement will be evidenced by correct formulae, nomenclature, mechanisms, significant figures, decimal places, units and other scientific terms and definitions.
- Candidates are expected to **use SI units** throughout; use of accepted derived units like L and mL are acceptable, but imperial units (oz, inches, F) and other units (e.g. cooking conventions) are not.
  - ❖ Accepted: ml or cm^3, L or dm^3.
  - ❖ Non-decimal system units (e.g. °F, cups and inches) are not appropriate.
- Bibliography: Any styles accepted by extended essays will be valid e.g. MLA style.

- Citations are expected for specific facts raised in the background or the conclusion.
- Incorrect use of terminology may include incorrect or missing units, repeated errors, inconsistent use of significant figures, incorrect use of terms. e.g. "my data proves..."
- A report outside of 1500 to 2250 words cannot receive full marks. Let's also discuss word count in detail since this is actually very important to follow.
- **Word count:** Teachers/Examiners would stop reading and stop awarding marks where the 2250 words limit has been exceeded. The communication criteria will be affected but possibly also others, depending on how much the script has over-run. Although there is no auto word count for IB examiners, they do count the words if they feel the report has gone above the limit.
- **Included in the word count:**
    - ❖ Main text: the context, analysis, and conclusion and evaluation, application wherever these appear.
    - ❖ Annotations & textboxes.
- **Not included in the word count:**
    - ❖ Title page.
    - ❖ Acknowledgments.
    - ❖ Contents page.
    - ❖ Titles and subtitles.
    - ❖ Citations, references and bibliography.
    - ❖ Footnotes—up to a maximum of 15 words.
    - ❖ Photographs.
    - ❖ Map legends and/or keys.
    - ❖ Labels—of 10 words or fewer.
    - ❖ Tables—of statistical or numerical data, or categories, classes or group names.
    - ❖ Calculations.

This sums up what you need to know about choosing a topic of IA and how to score well in an IA. Let's briefly go over some of the data analysis and gather methods commonly used.

4. HELPFUL TIPS

Surveys:

While making surveys make sure the questions and their options (if any) are designed in a way to provide differentiating information related to your variables. For example, while making a survey about EVS the options/questions must be able to differentiate between the 3 EVSs. This can be done by having 3 different options each from a different EVS.

What do you think should be done about deforestation?

  ◯ You believe that we can replace the role of trees in the environment with new technologies. The use of trees is much too essential in supporting the economy for us to stop consuming them.

  ◯ You believe that we should use more sustainable wood and paper products. They are an essential resource for producing things such as homes.

  ◯ You believe that we should majorly limit our paper and wood consumption. The planet need for trees is more important than ours.

Figure 3: A sample question for EVS based surveys.

The above sample question demonstrates how well-defined options can help easily distinguish between the 3 categories. It might be helpful for analysis that you may assign a numerical value to each option and then weigh total score to estimate the EVS for an individual. A score of 3 may be assigned to ecocentric option and 1 for technocentric option and the most frequent number or a high/low score may be used to assign EVS to individuals. More sample questions can be found here (https://www.surveymonkey.com/r/2JG8QH5) and a general guide on making surveys can be found here (https://www.sagepub.com/sites/default/files/upm-binaries/43589_8.pdf). For making and distributing surveys you may use Google forms or SurveyMonkey.

Secondary Data:

As discussed earlier we can use secondary data for an IA as well. However, these data must be from a reliable source and must be analysed by the student themselves. It is not recommended to use graphs from these secondary sources and rely completely on them as this takes away the analysis component. You should gather data but process everything on your own. In addition to acquiring data it is also recommended to highlight how this data was collected by the secondary source and what limitations it may have. Make sure to highlight time period or areas or any other parameters that have impacted this data collection. Few reliable sources are listed below:

Data Source	Web address
World bank	https://data.worldbank.org/
Our world	https://ourworldindata.org/
International Labour Organisation	https://www.ilo.org/global/statistics-and-databases/lang--en/index.htm
Food and Agriculture Organisation	http://www.fao.org/statistics/en/
UNESCO Institute of Statistics	http://uis.unesco.org/
Global Invasive Species Database	http://www.iucngisd.org/gisd/
International Union for Conservation of Nature and Natural Resources	https://www.iucnredlist.org/resources/summary-statistics#Summary%20Tables
Environmental Data Initiative	https://portal.edirepository.org/nis/home.jsp
World Wildlife Fund	https://www.worldwildlife.org/pages/conservation-science-data-and-tools
Pangaea	https://www.pangaea.de/
THE DIRECTORY OF OPEN ACCESS JOURNALS	https://doaj.org/

Table 1: Helpful reliable data sources.

Graphs and tables:

It is highly recommended to plot clear labelled figures. These should include units and correct data. Include a caption with ever figure that you draw. Although it is easy to draw tables in Microsoft word or Google docs, it is recommended to use dedicated specialised software for plotting graphs. Microsoft excel is an easy option; a guide on using this can be found here (https://www.biologyforlife.com/graphing-with-excel.html and https://www.biologyforlife.com/data-tables.html). A better option is GraphPad Prism (https://www.graphpad.com/scientific-software/prism/) This is a scientific statistical tool which Not only plots your data but can be helpful in analysing it as well. It has many statistical tools built in and can be used with a few clicks. A helpful tutorial can be found here (https://www.youtube.com/watch?v=gOwDNZoRNrw&ab_channel=GraphPadSoftware). Other tools include SOFA, SciLab, and JASP. A sample graph set with a sample caption made in GraphPad Prism is shown below:

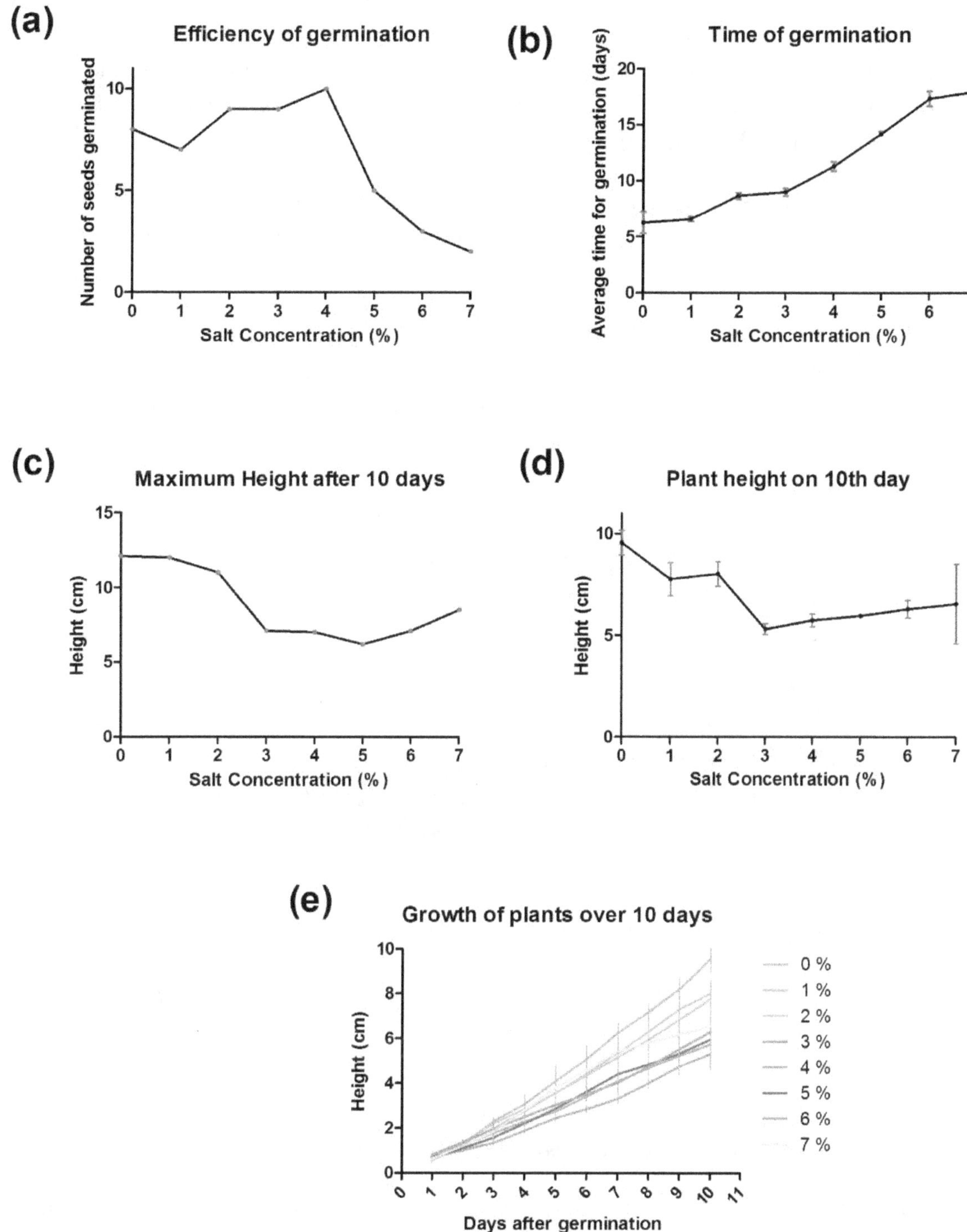

Figure 4: (a) Shows the efficiency of germination for each salt concentration, calculated from number of seeds germinated out of a total of 10 seeds in a 30 day period. (b) Average time (days) of germination for each salt concentration calculated by taking the average of the seeds that germinated. Error bars indicate Standard Error of Mean. (c) Maximum heath achieved by a single plant after 10 days for each salt concentration. (d) Average height of plants on the 10th day for each salt concentration. Error bars indicate Standard Error of Mean. (e) Shows the growth of plants for each salt concentration over a period of 10 days after germination. Each data point represents the average height of plants at that time point. Error bars indicate Standard Error of Mean.

Statistical Analysis:

Data processing is an important aspect of the IA. Here let's look at few common processing methods.

Mean: A mean is essentially the average value of a data set. This can be calculated by adding all the values and dividing by the number of values. It is recommended that you have at least 3 values for each measurement of which you want to take the average of. There are built in formulas/commands for this calculation in most software. You should not calculate means of already averaged values or of no-linear value (such as pH, which is logarithmic value).

Standard deviation (S.D.): Mean does not tell you a lot of the nature of a data set, particularly large data sets. To have a measure of spread of data standard deviation is necessary. A quick guide on S.D. can be found here

(https://www.youtube.com/watch?v=MRqtXL2WX2M&ab_channel=JeremyJones). GraphPad usually automatically calculates standard deviation and standard error (https://www.youtube.com/watch?v=A82brFpdr9g) for all data sets. In addition to calculating these values it would also be helpful to plot a box diagram for better visual comparison. A sample is given bellow:

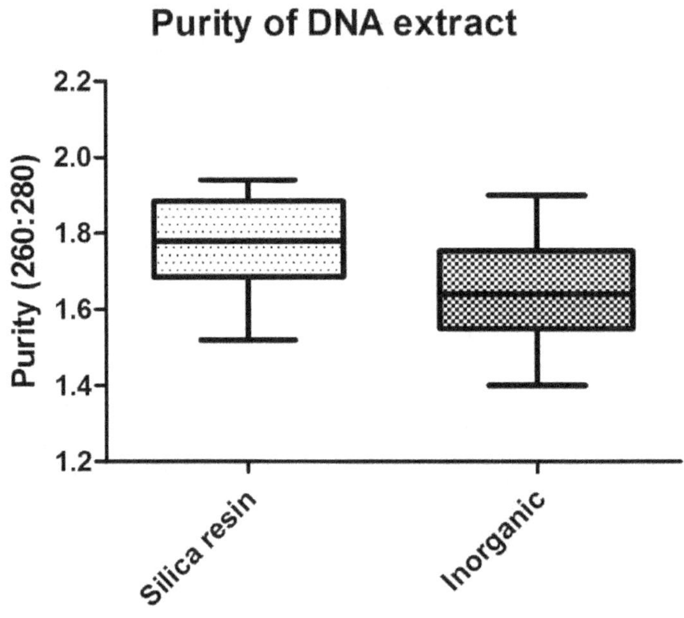

Figure 5: A sample box plot to show spread of data.

T-test and Chi-Square: A t-test should be use when you are comparing 2 data sets and want to know if they are significantly different or not. This helps you decide if the two values are different due to random chance or is that different due to other non-random factors. This type of analysis can be used for pre-treatment vs post-treatment or control vs experimental groups comparison. The test gives you a 'p-value'. This is an estimate of getting a difference in values due to chance. A p-value of 0.35 would mean there is a 35% chance the difference was just due to random chance. A p-value of less than 0.05 is usually considered significant and the difference is considered legitimate and not by chance. If you are comparing the same subjects (plants, humans etc) in two scenarios (before/after) then a 'paired t-test' is used and if these are two different groups altogether then an 'unpaired t-test' is used. In case more than 2 groups you may have to compare all of these with each other, this can be done with ANOVA.

If instead of comparing values between two or more measurements, you want to know if your experimental measurements follow a certain expected hypothesis you should perform a Chi-Square test. This compares the actual experimental value from a data set with an expected value based on a hypothesis. We use this method to see if the differences we observe are significant to reject a hypothesis.

A p-value of less than 0.05 would indicate a significant difference and you can sure this difference between observed and expected values is not due to chance and that the hypothesis must be inaccurate.

Bibliography:

1- Rutherford, Jill, and Gillian Williams. Environmental Systems and Societies Skills and Practice. Oxford University Press, 2016.
2- Emery, Nick. "Global Meets Local in the Ideal Structure." Campaign US, Campaign, 18 Feb. 2015, www.campaignlive.com/article/global-meets-local-ideal-structure/1321794.
3- "IB Animal Experimentation Policy." Extended Essay at ISB, ee.isb.ac.th/official-ib-resources/ib-animal-experimentation-policy.
4- Environmental/Social Impact Assessment and Permitting, aecom.com/nz/environmentalsocial-impact-assessment-permitting/.
5- "ESS Internal Assessments." *AMAZING WORLD OF SCIENCE WITH MR. GREEN*, www.mrgscience.com/ess-internal-assessments.html.
6- "Investigation Skills." BIOLOGY FOR LIFE, www.biologyforlife.com/investigation-skills.html.
7- *Environmental Systems and Societies Guide*. International Baccalaureate Organization, 2017.

PART II
SEVEN EXAMPLES OF EXCELLENT INTERNAL ASSESSMENT

The IAs featured in this section are all recently submitted assessments that scored exceptionally well after being moderated by the IBO. To prevent plagiarism and duplication of results, the appendices have been omitted. The IAs are presented in the exact same way as they were submitted, and without any edits or changes to formatting.

We do not retain the copyright of these commentaries, nor is this publication endorsed by the IBO. The Internal Assessments are being re-printed with the permission of the original authors.

1. TO WHAT EXTENT DOES EFFLUENT FROM PRINTING AND TEXTILE INDUSTRIES OF SANGANER AFFECT THE GERMINATION OF WHEAT SEEDS?

Author: Shreyan Jain
Moderated Mark: 30/30

Progress and prosperity are the two pillars of human happiness but one of the major problems we are facing in the 21st century is of the water pollution. We all know that water is essential for us, still we pollute it anyway. On the other hand, in the technological driven era, industrialization has led to many grave environmental concerns, challenging all living creatures.

Life began in water several million years ago, yet we are not concerned for the most precious resource i.e. water. Untreated wastewater is released in huge amount into the lakes, rivers and oceans. This problem has reached alarming proportions which is affecting us through different ways. We have finite sources of drinking water and if we look at the statistical data released by National Geographic; it points out, "About 70% of the Earth's surface is covered by water where only 2.5% is considered as fresh water. The remaining is ocean-based and saline. Moving on, only 1% of freshwater is accessible for humans, where most of it is locked up in snowfields or glaciers."[1]

In developing countries, most of the industries dispose their wastes either in water or on land without any treatment; the continuous discharging of effluent into the water bodies increases their toxicity level and affects aquatic life. Amongst the different types of pollution, water pollution caused by the discharging of effluent from industries causes severe problems not only to aquatic life but also to human life indirectly or directly.

In Sanganer area of Jaipur where I reside, there is a hub of small-scale print and textile industries. I have observed that the waste water from these industries of Sanganer is released into the local water body and in nearby areas. Farmers living in its proximity grow their crops using untreated waste water from printing or textile industries without any guilt as there is shortage of water in the region.

All the seasonal vegetables growing with polluted water is sold in the vegetable market of my locality. So, this made me curious to know that does this water affect the germination of seeds or the growth of plants. To learn more about it I thought of investigating on this through a research question that is **To what extent effluent from printing and textile industries of Sanganer affect the germination of wheat seeds.**

Experimental variables:

	Named Variables	Units	Equipment or Procedure for measurement/Control	Justification
Independent	Amount of effluent to be added.	Milliliter (ml)	Measuring cylinder will be used to measure the amount of effluent to be added in water and for watering the plant under investigation.	Use of measuring cylinder will help in using precise amount of effluent to be added in water.
Dependent	Number of wheat seeds germinated.	-	By manually counting the number of germinated seeds in every pot.	Because this is the easiest and convenient method and does not incur any cost.

---

[1] https://www.nationalgeographic.com/environment/freshwater/freshwater-crisis/

Controlled	Amount of water	Milliliter (ml)	Measuring cylinder will be used to measure the amount of water.	Measuring cylinder gives accurate results, it is easy to handle. Same measuring cylinder will reduce the errors.
	Named Variables	**Units**	**Equipment or Procedure for measurement/Control**	**Justification**
Controlled	Type of soil	-	Same type of soil i.e. loam will be used for all the experimental pots.	Because different types of soil have different water holding capacity and the amount of nutrient which may affect the germination of wheat seeds.
	Amount of soil	grams	To measure the amount of soil to be put in each pot digital weighing machine will be used.	Digital weighing machine helps in accurate measurement as difference in amount of soil may affect the germination of seeds.
	Number of seeds	-	15 seeds will be sown in each pot by counting manually.	If different number of seeds will be sown then due to change in amount of space between the seeds, amount of water they will receive, etc. may affect the germination.
	The shape and size of the pot.	Centimeter (cm)	Pot of same shape and size of diameter 15cm will be used in the experiment.	Different size and shape of the pot will have different surface area exposed to environment which could influence the germination of the seeds.
	Day of noting the observation.	-	Number of seeds germinated will be counted after same number of days, here it is 4 days.	Collecting the data of number of seeds germinated on different days may give improper results as the number of germinated wheat seeds will differ from each other on different days.

Safety/ethical considerations:

One needs to handle untreated effluent with caution and use gloves for the experiment; as it is a toxic substance.

Wear safety goggles while making the concentration as effluent may enter your eyes during the process.

Keep the effluent in a safe place where it does not come in contact with anyone and dispose the soil after the experiment in a proper manner as it contains untreated effluent.

Equipment list or material

Pots: 25 pots of same size and shape

Effluent from printing/textile industry: 400 ml

Measuring cylinders (100 ml): 1

Measuring cylinders (10 ml): 1

Beakers (1L): 5

Loamy soil: 5.5 kg

Wheat seeds: 450 seeds

Tap Water: 3.6 L

Labels: 25

Marker: 1

Digital weighing machine: 1

Graphic Display Calculator: 1

Methodology

Take 25 pots. Mark the first 5 pots as A1 to A5 then B1 to B5 then C1 to C5 then D1 to D5 and last 5 as E1 to E5.

Soak all the wheat seeds in water for ten minutes and remove the floating seeds as they are non-viable seeds, so only the viable seeds should be sown as to reduce the error.

Fill all the pots with 200 grams of loamy soil and sow 15 wheat seeds in each with sufficient spacing and at proper depth.

Keep all the pots at same place in open area so that they receive same environmental conditions.

Sprinkle 20 ml of solution on all the pots two times in a day for 4 days, once 7am in the morning and next at 5pm in the evening to keep the soil moist for germination as mentioned below-

Water Pot A1-A5 with solution containing 0% effluent.

Water Pot B1-B5 with solution containing 5% effluent.

Water Pot C1-C5 with solution containing 10% effluent.

Water Pot D1-D5 with solution containing 15% effluent.

Water Pot E1-E5 with solution containing 20% effluent.

Count the number of seeds germinated in each pot after 4 days.

Note down the readings.

Analyze the data.

Table 1. Shows the number of germinated seeds in the pot given water with no effluent i.e. 0% effluent.

S. no	Pot	Number of seeds sown	Number of seeds germinated
1	A1	15	15
2	A2	15	14
3	A3	15	15
4	A4	15	14
5	A5	15	14

Table 2. Shows the number of germinated seeds in the pot watered with 5% effluent.

S. no	Pot	Number of seeds sown	Number of seeds germinated
1	B1	15	13
2	B2	15	13
3	B3	15	14
4	B4	15	13
5	B5	15	12

Table 3. Shows the number of germinated seeds in the pot watered with 10% effluent.

S. no	Pot	Number of seeds sown	Number of seeds germinated
1	C1	15	12
2	C2	15	11
3	C3	15	12
4	C4	15	13
5	C5	15	11

Table 4. Shows the number of germinated seeds in the pot watered with 15% effluent.

S. no	Pot	Number of seeds sown	Number of seeds germinated
1	D1	15	10
2	D2	15	10

3	D3	15	11
4	D4	15	9
5	D5	15	11

Table 5. Shows the number of germinated seeds in the pot given 20% effluent.

S. no	Pot	Number of seeds sown	Number of seeds germinated
1	E1	15	10
2	E2	15	9
3	E3	15	9
4	E4	15	8
5	E5	15	10

Data Processing

Formula used

$$average\ number\ of\ seeds\ germinated = \frac{total\ number\ of\ seeds\ germinated\ in\ all\ the\ pots\ watered\ with\ same\ \%\ of\ effluent}{5}$$

Sample calculation for 5% effluent

Number of seeds germinated in pots **B1 to B5** with **5% effluent**

$$average\ number\ of\ seeds\ germinated = \frac{13 + 13 + 14 + 13 + 12}{5} = \frac{65}{5} = 13$$

Table 6: Shows the average number of seeds germinated in different types of pots watered with different % of effluent

Effluent added (%)	Average number of seeds germinated
0	14
5	13
10	12
15	10
20	9

Formula used

$$\% \ of \ seeds \ germinated = \frac{Total \ number \ of \ seeds \ germinated \ in \ all \ the \ pots \ watered \ with \ same \ \% \ effluent}{Total \ number \ of \ seeds \ sown} \times 100$$

$$\% \ of \ seeds \ germinated \ in \ Pot \ B = \frac{65}{75} \times 100 = 86.6667\% = 86.7\%$$

Table 7: Shows the % of wheat seeds germinated when watered with water having different concentration of effluent.

S. no	Pot type	% of effluent in water	% of germinated seeds
1	A	0	96.0
2	B	5	86.7
3	C	10	78.7
4	D	15	68.0
5	E	20	61.3

Graph 1: Shows % of wheat seeds germinated with water of different % of effluent.

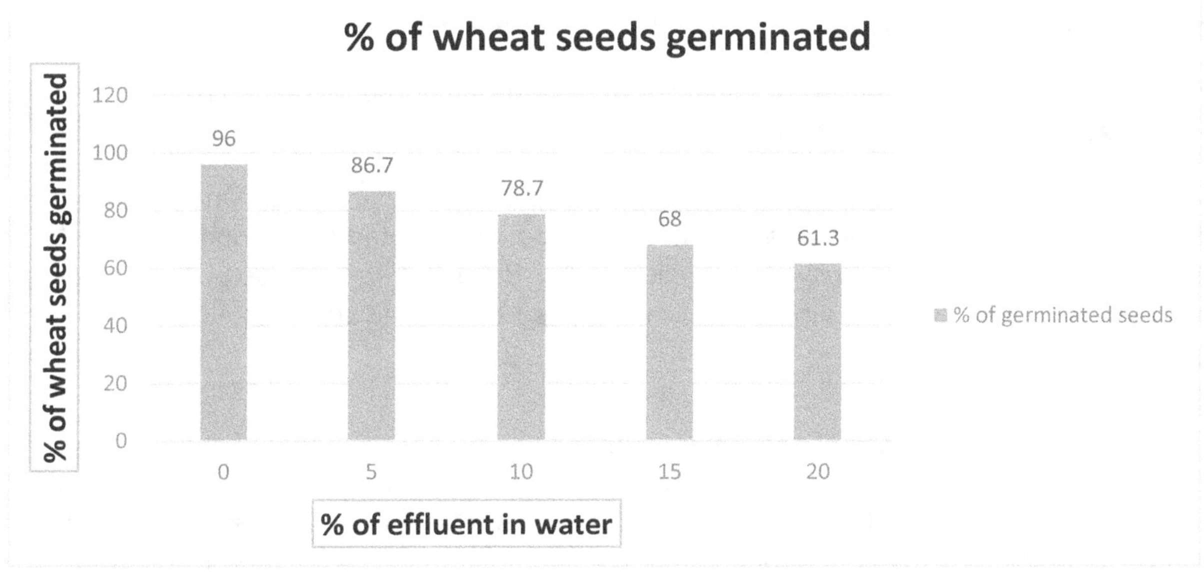

Statistical Analysis

To further validate the result, statistical analysis was done. For this Pearson's correlation coefficient was calculated by **TI-*n*spire GDC**. The value of **r** came to be   -0.99819, which is very close to -1, this demonstrates a very strong negative correlation. Since the value of r is negative, both the variables are inversely correlated. This is evident that as the % of effluent increases, the % of germination of wheat seeds decreases.

Discussion and analysis

Table 7, 8 and the graph 1 gives the impression that industrial effluent has a strong negative influence on the germination of wheat seeds. As the concentration of effluent increases, % germination of wheat seeds decreases. When the wheat seeds were given water with 0% effluent i.e. Pot A type, then the % germination was seen to be the maximum of 96%. Whereas, when seeds were given water with 5% effluent then only 86.7% of wheat seeds germinated.

The presence of toxic metals in effluent such as lead, copper, nickel, zinc, etc. when comes in contact with soil, reduces the quantity and quality of food by preventing the uptake of essential nutrients which further affects the germination of seeds and growth of plants.

So, this possesses a threat to the environment and is hazardous to human beings too. The untreated effluent is released continuously which affects the agriculture sector and also has several diverse effects on water. It does not only affect the germination of seeds but also affects the growth of the plants, where the length of the crop, the weight of the crop and the amount of nutrients in it decreases due to toxic chemicals.

As per the U.S EPA, everyday 3.28 billion gallons of household and industrial sewage are being dumped in the nation's water which pollutes the water resources[2]. Various nations across the globe are suffering from the problem of water scarcity, this means that clean water is vital. The discharged wastewater can seep into underground water resources and pollute it, thus making it unsuitable for the upcoming generation.

As percentage of germination decreases due to the increase in concentration of effluent which is released via industries either makes the soil too acidic or too alkaline, it also contains toxic elements which affects the soil by washing away the essential nutrients it has and also killing the microorganisms living in it. There are many industries worldwide which are continuously releasing or disposing of their waste directly into the water bodies and causing a harm to the environment where plants, animals and humans are suffering due to their direct or indirect impact.

The untreated effluent also contains wide variety of **pollutants** and host disease-causing microbes which is the reason behind the waterborne diseases.

But, the problem of water pollution is increasing day by day and affecting the humans as well as the environment. The available freshwater on our earth is getting polluted thus reducing its availability. This leads to water scarcity/stress which further forces humans to use polluted water for various purposes and leading to a problem for which we need to find solutions.

Conclusion and its Evaluation

Through the results of the experiment and from the data it can be analyzed that as the concentration of the effluent increases from 0% to 20%, the percentage of germination of wheat seeds decreases which is evident that polluted water is not suitable for development and growth of plants. Industries create employment opportunities in a country and increase its GDP but they need to take measures to protect the environment so they should not release effluent directly into the water bodies/land as it causes negative impacts on human and aquatic life.

Strength of the method

The experiment was repeated 5 times to increase the accuracy and authenticity of the result.

The method used was easy and convenient and the apparatus taken into consideration were giving accurate measurements to minimize errors.

The possible variables were tried to kept controlled in the best possible manner as mentioned earlier.

---

[2] https://www.organicawater.com/effects-wastewater-environment/

The experiment was not conducted in the natural environment so all the natural conditions were not fulfilled.

The result may vary if the experiment is conducted with effluent from different industries. As the effluent from various resources may have a different chemical composition from each other.

Scope of improvement/ Further area of research

We could consider effluent from different sources because it may vary in concentration.

Different types of soil and a greater number of seeds could be investigated to get more accurate and authentic results.

Apart from counting the number of seeds germinated, we can also measure the length of the seedlings which will make it more interesting and precise.

Possible Solutions

Strict laws should be made for industries to treat wastewater before releasing them into water bodies.

In order to control the release of untreated effluent into the environment, the government should impose some strict laws against it and monitor them properly. Effluent which has the pollutants more than the permissible range has to be treated before releasing into the water bodies/the environment. If industries will treat the effluent before releasing it then the pollution rate will decrease at a large scale and benefit the agricultural sector and human life as we will become less prone to diseases being caused by the release of effluent which pollutes our surroundings.

Strength

As the strict laws will be in action and if monitored properly then it will reduce the problem of releasing harmful pollutants to a large extent. Though it's an anthropocentric way to solve the issue but to control the situation with the effect it has to be initiated in this way and side by side eco-centric way by creating awareness and technocentric way by cleaning up the water bodies to restore them should go along with it.

Enforcement of strict laws and fines will pressurize the industry owners to treat wastewater before releasing it into the environment.

It will not only help in maintaining ecosystem health and biodiversity but indirectly will reduce the frequency of the toxic substances entering the food chain. The agricultural sector and aquatic life would also be benefitted as the treated effluent is not so harmful as the untreated effluent.

Weakness and limitations

It would be difficult to monitor the small-scale industries.

The treatment of effluent is not cost-effective so the owners of small industries may not be able to afford the treatment.

Sometimes the political factors also refrain the government to implement strict laws as the democratic government has fear of loss of vote banks.

2. THE EFFECT OF PH ON DISSOLVED OXYGEN IN WATER

Author: Katheryn Weiler
Moderated Mark: 25/30

Background Research/Environmental Impacts.

Ocean deoxygenation is the loss of oxygen in a body of water due to anthropomorphic impacts, mainly the output of CO_2. The output of CO_2 is becoming a bigger issue in our world as time has continued to go on due to the fuel burning industry expanding as human demand increases. Not only does it have a severe impact on global warming as CO_2 absorbs long wavelengths reflected by the earth's surface, hence keeping them in the atmosphere and trapping the heat.

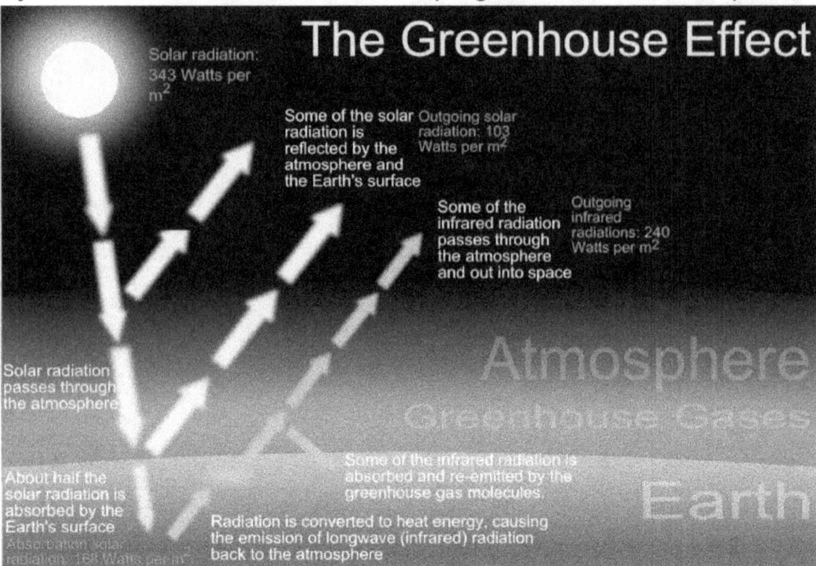

Figure One: Example of Global Warming (23)

However, it is not known the effect that the extra CO_2 has on the ocean. The ocean is one of the largest carbon sinks in our planet. It absorbs roughly 25% of the CO_2 in the atmosphere.(2) The CO_2 in the oceans reacts with the water in our ocean to form carbonic acid ($CO_2 + H_2O \rightleftharpoons H_2CO_3$). The carbonic acid then dissociates to form bicarbonate ions ($HCO_3^-)^+$, which then further breaks down to carbonate ions (CO_3^-) and hydrogen ions (H^+).

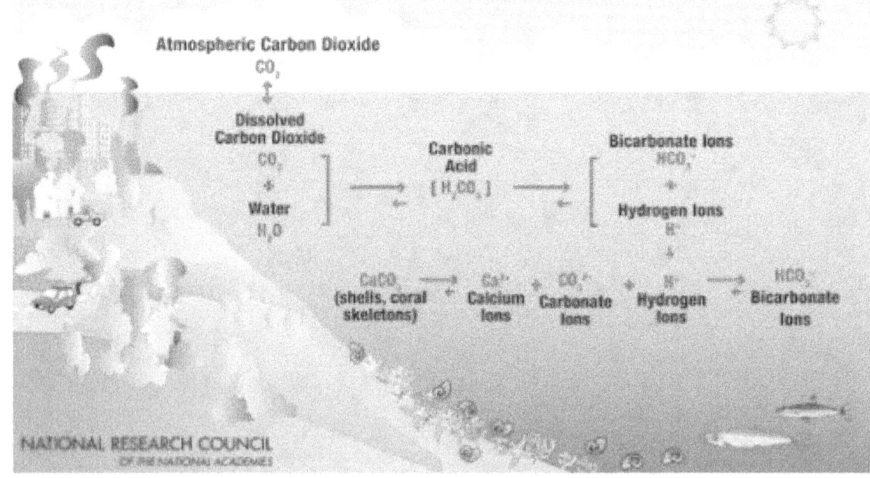

Figure Two: The Process of Ocean Acidification (11)

While carbonate is used by many organisms, including crustaceans who use it to make calcium carbonate shells for protection, too much of it can be deadly to the organisms that use them. (1)For example, an organism like clams use it to form hard shells to stay protected.So while it is essential to have dissolved CO_2 in the oceans for the purpose of photosynthesis of aquatic plants, excessive concentration of CO_2 is harmful because of the Hydrogen Ions produced. The concentration of hydrogen ions is what determines the pH. (3) Since the beginning of the industrial revolution, the pH in the oceans has fallen by 0.1 pH. That may not seem like a lot, but it is a 30% increase in acidity due to the pH scale being logarithmic. This is already having a severe effect on aquatic life. For example, despite there being more calcium carbonate due to more CO_2 in the water, the acidity of the ocean is causing a faster breaking down of organisms shells, which are crucial for their survival. It is also destroying aquatic habitats. For example,the more alkaline water can damage coral.(4) Elkhorn coral, a coral the originates from the caribbean, is said to have their fertilization inhibited due to ocean acidification and is currently classified as "Critically Endangered" on the IUCN red list. (5) Coral is an important part of any saltwater aquatic ecosystem, as they are used for shelter by aquatic creatures as well as providing food to others.

Despite the many consequences of increased acidity and pH, the biggest issue is associated with a decrease in dissolved oxygen levels. Dissolved oxygen is one of the many factors that are looked at when analyzing the health of an aquatic system. There have been many studies conducted over the years, including one done by Changjuan Zang and other associates done in 2010, analyzing relationships between pH and dissolved oxygen over two decades. They came to the conclusion that there is a significant positive linear correlation between dissolved oxygen and pH levels. Essentially, as pH goes up, making the water more basic, there is a rise in O_2 levels. And inversely, as the pH drops, making the water more acidic, the dissolved oxygen levels drop. (6) Dissolved oxygen is one of the most important components of aquatic life. A lower level of dissolved oxygen can have an increased stress on aquatic life. The recommended amount of oxygen for fish to breath through their gills is 5mg/L of dissolved oxygen. Fish start to get distressed between 2-4 mg/L. When it drops below 2mg/L, this is where death starts to occur. (7) As you can see in the figure below, around 7ppm is where we can see a support in growth for fish populations and at 3-5 ppm it becomes very stressful for fish and they will not survive for long.

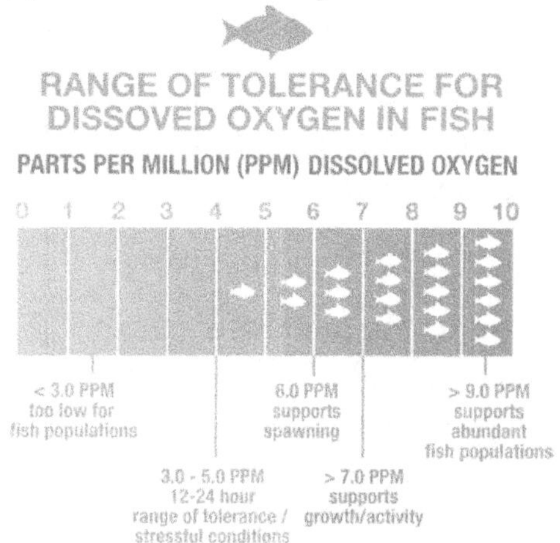

Figure Three: A Range of Dissolved Oxygen's Effect on Fish. (24)

Knowing that there is currently a lot of debate about Global Warming and its impacts on the environment, that inspired be to come up with a research question that focuses on other effects of global warming.I came up with the research question;To what extent does the pH of a buffer added to water affect the level of dissolved Oxygen using the Winkler titration method? The issue of global warming connects to my research question about dissolved oxygen because the more CO_2 trapped in the atmosphere, the more it gets dissolved into the bodies of water, lowering the pH and causing a decrease in dissolved oxygen.

To analyze the effect of pH and dissolved oxygen, I turned to the method that had got me interested in this lab in the first place. The Winkler titration was developed by Ludwig Wilhelm Winkler, in 1888. The key to this experiment is that the amount of iodine produced in proportional to the amount of oxygen in the water. (8) I chose the method of using the Winkler Titration as my method due to the prior knowledge and experience. I do not know how to effectively use a dissolved oxygen probe. I am not experienced with it, but I am experienced with using titrations and stoichiometric relationships to calculate values that I need.

Chemical Equations of the Winkler Titration Reaction:

1. $MnSO_4 + 2\ KOH \dashrightarrow K_2SO_4 + Mn(OH)_2$
2. $2\ Mn(OH)_2 + O_2 \dashrightarrow 2\ MnO(OH)_2$
3. $MnO(OH)_2 + 2\ H_2SO_4 \dashrightarrow 3\ H_2O + Mn(SO_4)_2$
4. $Mn(SO_4)_2 + 2\ KI \dashrightarrow K_2SO_4 + MnSO_4 + I_2$
5. $2\ Na_2S_2O_3 + I_2 \dashrightarrow Na_2S_4O_6 + 2\ NaI$

(20)

<table>
<tr><td>

Variables:

</td></tr>
<tr><td>

Independent Variable: My independent variable is the pH of the buffer that is combined with the water. For this manipulation, I will be using four different buffers with a pH of 2,4,7, 9, and 12 with 100ml of each.

</td></tr>
<tr><td>

Dependent Variable: My dependent variable is the volume of Sodium Thiosulfate needed to change the color of the water sample. Knowing this, I can find the Parts per Million (ppm)of dissolved oxygen using the ratio of the balanced equations. It is a 4:1 ratio.

</td></tr>
<tr><td>

Controlled Variables:
Same ratio of Water to Buffer:
The more volume of buffer there is, the higher the concentration of hydrogen ions. Because we only want to look at the pH of the buffer, it is important to have the same volume of buffer each time. To control this, I will add 100ml of buffer solution to 200ml of water to create a 300ml solution,keeping the ratio of buffer to water as a 2:1 ratio.
Same volume of water:
In simple terms, the more water used in each trial will have an effect on the amount of iodine. This is simply due to there being more in the volume, thus it would take more $Na_2S_2O_3$ to get NaI. This would give inconsistent results as we only want to see how the pH of the buffer affects the dissolved oxygen. To control this, each trial will only be used with 50ml of water, which will be measured using a 25ml pipette,meaning two full pipettes of water per trial.
Same concentration of iodine
Because the concentration of iodine is directly related to the amount of dissolved oxygen, each trial needs to have the same concentration of iodine inserted into the solution. To keep this consistent, each 300ml of water will have 2ml of alkaline iodide (1 mol dm^3 of KOH and 1 mol dm^3 KI). This ensures that the only effect on dissolved oxygen is the pH.
Control of extra O_2 in bottle:
One of the most difficult things to control in this experiment is making sure that there is no excess oxygen in the bottle with the water. This is one of the biggest limitations in the design of the winkler method. To reduce this limitation as much as possible, I will avoid opening the bottle unless needed to obtain water or add any of the chemicals. All the water will be kept in the same type of bottle with a lid. The bottle I used are normally used for Vernier Oxygen Probes, but works very well for this type of experiment.

</td></tr>
</table>

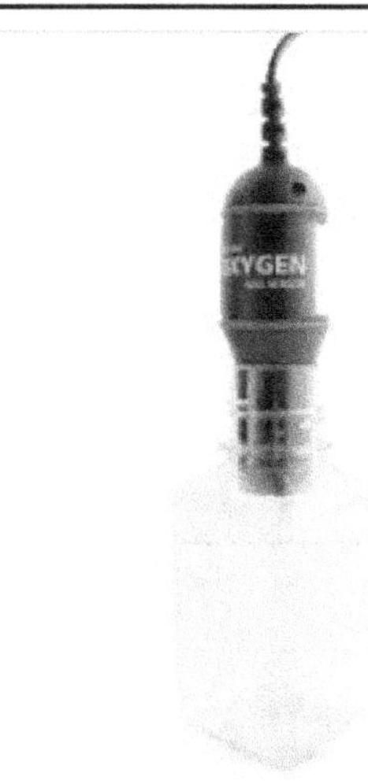

Figure Two: The bottle that I will be using.(10)

Same water source:

The concentration of other dissolved ions have a significant effect on the dissolved oxygen. For example, distilled water has no ions in it while salt water contains many more dissolved Sodium and Chlorine ions. To keep this consistent, all the water that was tested was from the schools tap water faucet, which all comes from the same source. To keep it even more consistent, it will use the same faucet each time.

Same Temperature of Water

Temperature has an extreme effect on dissolved oxygen. The warmer the water, the more dissolved oxygen in the water sample. To keep all the water samples consistent, they are all going to be at room temperature (25°C). This will be monitored with a thermometer before any data is collected

Materials.

-     50 ml Burette
-     Burette clamp
-     Ring stand.
-     Paper Towel
-     Magnetic Stirrer.
-     Funnel
-     1 ml plastic pipette
-     3 x 10 ml pipettes
-     25 ml pipette
-     250 ml erlenmeyer flask
-     5 x 300 ml BOD bottle
-     3 ml pipette

-   250 ml of 0.002 M $Na_2S_2O_3$
-   10 ml of Alkaline Iodide
-   10ml of 0.5M Manganese Sulfate
-   10 ml of Concentrated Sulphuric Acid
-   1% Starch Solution
-   100 ml of pH 2 Buffer
-   100 ml of pH 4 Buffer
-   100ml of pH 7 Buffer

-   1 L of Tap Water
-   100 ml of pH 9 Buffer
-   100 ml of pH 12 Buffer

Safety:

Concentrated Sulphuric Acid:
Sulphuric Acid in this form is extremely corrosive. It can cause burns if it comes in contact with skin. Because it is concentrated it can cause second or third degree burns. If it comes in contact with skin, immediately wash for a minimum of 20 minutes. If it is on clothing immediately remove and seek immediate medical attention. If it gets into the eyes, it can cause corneal damage, which could lead to permanent blindness. If this occurs, immediately flush for 20 minutes, seek medical attention immediately. This product should not be ingested under any circumstances, but if it is, then do not induce vomiting. If the person is awake and lucid, give 1/2 to a glass of water to dilute solution. Contact a local poison center immediately. (12)

Sodium Thiosulphate:
Because the concentration is very low, it doesn't cause a huge potential threat. It should not be ingested under any circumstances. If it is ingested, do not induce vomiting. Loosen tight clothing and seek medical help. If it gets on skin, wash with water and make sure that the area doesn't become more infected. If it gets in eyes, wash immediately for 15 minutes. If irritation continues, seek medical attention. (13)

Alkaline Iodide:
Alkaline Iodide can be quite dangerous for this experiment. If it comes into contact with the eyes, wash immediately for 15 minutes and seek medical attention. If it comes into contact with skin, wash area for 15 minutes and seek medical attention. If it is ingested, do not induce vomiting. If the person in conscious, have them drink water. Seek medical attention.(14)

Manganese Sulfate:

If this chemical comes into contact with the eyes, wash immediately for 15 minutes and seek medical attention. If it comes into contact with skin, wash area for 15 minutes and seek medical attention. If it is ingested, do not induce vomiting. If the person in conscious, have them drink water. Seek medical attention.[15]

Buffer Solutions:
If this chemical comes into contact with the eyes, wash immediately for 15 minutes and seek medical attention and contact a poison control center. If it comes into contact with skin, wash area for 15 minutes and seek medical attention and poison control center. If it is ingested, do not induce vomiting. If the person in conscious, have them rise the mouth with water. Seek medical attention and contact a poison control center. [16,17,18,19]

Ethical Concerns:
This project does not involve the use of any animal participants or any living creatures. Thus, ethical concerns are not an issue here.

Environmental Concerns
All of the chemicals were disposed of properly based on the MSD sheets.

Procedure:

1. Obtain safety goggles.

2. Fill one Biological Oxygen Demand (BOD) bottle with 100 ml of pH 2 buffer.

3. Add 200 ml of tap water to the bottle.
    a. Continue filling until the bottle overflows slightly and there are no clear air bubbles at the top.

4. Using a 10 ml pipette, add 2 ml of 0.5M Manganese Sulfate ($MnSO_4$)
    a. Make sure that the tip of the pipette is submerged into the water

5. Using a different 10 ml pipette, add 2 ml of Alkali Iodide (1 mol dm^3 of KOH and 1 mol dm^3 KI)
    a. This should cause a brown/yellow precipitate to form.

6. Wait for the precipitate to settle at the bottom of the bottle.

7. Once the precipitate has settled, add 2 ml of Concentrated Sulphuric Acid (H_2SO_4)
    a. Adding this should dissolve the precipitate
    b. Wipe around the bottle to ensure that there is no sulphuric acid when you touch it.

8. Close top and shake a few times until the precipitate is fully dissolved.

9. Set up your burette.
    a. Clip the burette holder onto the ring stand.

b. Using a funnel, fill up the burette with 50 ml of 0.002M Sodium Thiosulphate $(Na_2S_2O_3)$
c. Place the magnetic stirrer base underneath the tip of the burette.
d. Place a piece of white paper towel on top so that it is easy to see the color change.
e. Place the magnetic pill into a 250 ml erlenmeyer flask and then place the flask on top of the stirrer base.

(This is what the setup should look like)

10. Using the 25 ml pipette, add 50 ml of your water solution to the flask.
    a. Close the bottle each time you are done collecting your sample.
    b. Take note of the starting value on the burette

11. Turn on the magnetic stirrer and start titrating. Stop the titration when your water becomes a pale straw color.
    a. Write down the change in volume in the burette

12. When that color is desired, add a few drops of the starch solution until the solution in the flask becomes a deep purple.
    a. Take note of your starting volume on the burette.

13. Titrate the solution until it is a clear, colorless liquid.
    a. Take note of the final volume on the burette.

14. Dump out the contents of the flask and be sure to wash the flask and the magnetic stirrer thoroughly.

15. Repeat steps 10-14 4 more times to give you a total of 5 trials per bottle.

16. Repeat steps 2-8 and 10-15 4 more times with buffer solutions with a pH of 4,9, and 12 and once with no buffer solution and 300ml of water. Fill up the burette when needed, but keep track of the values. (20)

Raw Data

Table One: The Initial and Final Volume of Sodium Thiosulphate in a Burette when it is Titrated against 300ml of water with various pH's (2,4,7,9,and 12). The reaction is considered done when the starch in the solution went from purple to colorless.

pH of the Buffer	Volume in the burette (ml) (±0.05)	Trial One	Trial Two	Trial Three	Trial Four	Trial Five
12	Initial	50.00	42.60	34.80	27.20	21.70
	Final	42.60	34.80	27.20	21.70	13.00
9	Initial	50.00	47.20	44.80	42.30	39.80
	Final	47.20	44.80	42.30	39.80	37.00
7	Initial	50.00	24.60	50.00	27.90	21.10
	Final	24.60	1.00	27.90	7.70	0.00
4	Initial	37.50	35.80	34.10	32.00	30.30
	Final	35.80	34.10	32.40	30.30	28.60
2	Initial	28.60	27.70	26.90	25.50	24.00
	Final	27.70	26.90	25.50	24.00	22.80

Qualitative Observations:

(**First Photo:** Color of the water with a pH buffer of 2)
(**Second Photo:** Color of waters with (from left to right) pH 7, pH 12, and 9)

One of the most interesting parts of my investigation was the difference in color. As you can see in the photos, the less Sodium Thiosulphate used to titrate, the lighter the color of the water. When the Alkaline Iodide was added, all of the samples formed a precipitate, however, only the sample with no buffer formed a brown-yellow precipitate. The other samples formed a more yellow-white precipitate, with it getting lighter as the pH of the buffer dropped. The sulphuric acid immediately dissolved the precipitate as the Manganese displaced the hydrogen and it formed $Mn(SO_4)_2$. After the first round of titration,starch solution was added, which turned a deep purple. The same amount would be added to the solutions with a lower pH and it would be more clear, while the higher pH's would be more opaque.

Sample Calculations

Change in Volume in the Burette.
Final Volume (ml)-Initial Volume (ml)=Δ in Volume (ml)
Δ in Volume (ml)=25.40 (ml) ($\pm$0.05)-0.00 (ml) ($\pm$0.05)
Δ in Volume = 25.40 ml ($\pm$0.1)

Moles of Sodium Thiosulphate Used to Titrate:
Moles=volume (dm^3) x concentration (mol dm^{-3})
Moles of Sodium Thiosulphate = 0.0254 (dm^3) ($\pm$0.39%) * 0.002 mol dm^{-3})($\pm$5 %)
Moles of Sodium Thiosulphate= 1.48×10^{-5} moles ($\pm$5.39%).

Stoichiometric Ratio of Sodium Thiosulphate to Oxygen.
1. $MnSO_4 + 2\ KOH \rightarrow K_2SO_4 + Mn(OH)_2$
2. $2\ Mn(OH)_2 + O_2 \rightarrow 2\ MnO(OH)_2$
3. $MnO(OH)_2 + 2\ H_2SO_4 \rightarrow 3\ H_2O + Mn(SO_4)_2$
4. $Mn(SO_4)_2 + 2\ KI \rightarrow K_2SO_4 + MnSO_4 + I_2$
5. $2\ Na_2S_2O_3 + I_2 \rightarrow Na_2S_4O_6 + 2\ NaI$

Using the above equation's stoichiometric ratios, we can deduce that for every 2 moles of $Na_2S_2O_3$, we need 1 mole of Iodine. Thus we need to take the moles of Sodium Thiosulphate and divide it by 2 ($1.48 \times 10^{-5}/2$) to get the moles of I_2 (7.4×10^{-6}). Continuing to work backwards, we go to $Mn(SO_4)_2$, which is a 1:1 ratio. Then we go to $MnO(OH)_2$, which is still a 1:1 ration. Then we have a 2:1 ratio of $MnO(OH)_2$ to Oxygen. Thus we have to divide our moles of $MnO(OH)_2$ by 2 (7.4×10^{-6}m/2). That gives us 3.7×10^{-6}($\pm$21.56 %) moles of Oxygen gas.

Moles to Parts Per Million (ppm)
$$\frac{Mass\ of\ Solute}{Mass\ of\ Solution} \times 10^6 = Parts\ Per\ Million\ (ppm)$$
Mass of Solute (g)=Moles x Molar Mass
Mass of Solute (g)= 3.7×10^{-6}($\pm$21.56 %) moles x 32.00 g
Mass of Solute (g)=1.18×10^{-4} ($\pm$21.59 %)
$$\frac{1.18 \times 10{-4}\ g\ (\pm 21.59\ \%)}{50\ g\ (\pm 0.02\%)} \times 10^6 = Parts\ Per\ Million\ (ppm)$$
Parts Per Million (ppm)= 2.37 ppm ($\pm$21.61%)

Average Calculations
$$\frac{[Trial\ 1\ (Unit)+Trial\ 2\ (Unit)+Trial\ 3\ (Unit)+Trial\ 4\ (Unit)+Trial\ 5(Unit)]}{5} = Average\ for\ all\ Trials$$
$$\frac{[25.4\ (ml)+23.6\ (ml)+22.1\ (ml)+20.2\ (ml)+21.1(ml)])}{5} = Average\ Change\ in\ Volume\ (ml)\ (\pm 0.1)$$

Average Change in Volume =22.5 ml ($\pm$0.1)

Processed Data:
Table Two: The Change in Volume of Sodium Thiosulphate in a Burette when it is Titrated against a Solution of Water, Alkaline Iodide (KI+KOH), Manganese Sulfate ($MnSO_4$) and Concentrated

Sulphuric Acid (H_2SO_4). The reaction is considered done when the starch in the solution went from purple to colorless.

	Trial 1	Trial 2	Trial 3	Trial 4	Trial 5
pH of Buffer	Change of Volume in the Burette (ml) (± 0.1)				
12	7.4	2.8	25.4	1.7	0.9
9	7.8	2.4	23.6	1.7	0.8
7	7.6	2.5	22.1	1.7	1.4
4	6.1	2.5	20.2	1.7	1.5
2	8.1	2.8	21.1	1.7	1.2
Average	7.4	2.6	22.5	1.7	1.2

Table Three: The moles of Oxygen in Each Water Sample when it is Titrated against a Solution of Water, Alkaline Iodide (KI+KOH), Manganese Sulfate ($MnSO_4$) and Concentrated Sulphuric Acid (H_2SO_4). The reaction is considered done when the starch in the solution went from purple to colorless.

	Trial 1	Trial 2	Trial 3	Trial 4	Trial 5
pH of Buffer	Moles of O_2 in each water sample ($\pm 21.56\ \%$)				
12	3.70×10^{-6}	1.40×10^{-6}	1.27×10^{-5}	8.50×10^{-7}	4.50×10^{-7}
9	3.90×10^{-6}	1.20×10^{-6}	1.18×10^{-5}	8.50×10^{-7}	4.00×10^{-7}
7	3.80×10^{-6}	1.25×10^{-6}	1.11×10^{-5}	8.50×10^{-7}	7.00×10^{-7}
4	3.05×10^{-6}	1.25×10^{-6}	1.01×10^{-5}	8.50×10^{-7}	7.50×10^{-7}
2	4.05×10^{-6}	1.40×10^{-6}	1.06×10^{-5}	8.50×10^{-7}	6.00×10^{-7}
Average	3.7×10^{-6}	1.30×10^{-6}	1.13×10^{-5}	8.50×10^{-7}	6.00×10^{-7}

Table Four: The Average Parts Per Million of Oxygen in Each Water Sample when it is Titrated against a Solution of Water, Alkaline Iodide (KI+KOH), Manganese Sulfate ($MnSO_4$) and Concentrated Sulphuric Acid (H_2SO_4). The reaction is considered done when the starch in the solution went from purple to colorless.

pH of the Buffer Solution	Average Parts Per Million of Oxygen in each 50ml water sample. (ppm) ($\pm 21.61\%$)
2	0.38
4	0.54
7	7.23
9	0.83

Graph One: The Average Parts Per Million of Oxygen in Each Water Sample when it is Titrated against a Solution of Water, Alkaline Iodide (KI+KOH), Manganese Sulfate (MnSO$_4$) and Concentrated Sulphuric Acid (H$_2$SO$_4$). The reaction is considered done when the starch in the solution went from purple to colorless.

Equation of Trendline: $y = 0.383e^{0.1727x}$

$R^2 = 0.32391$

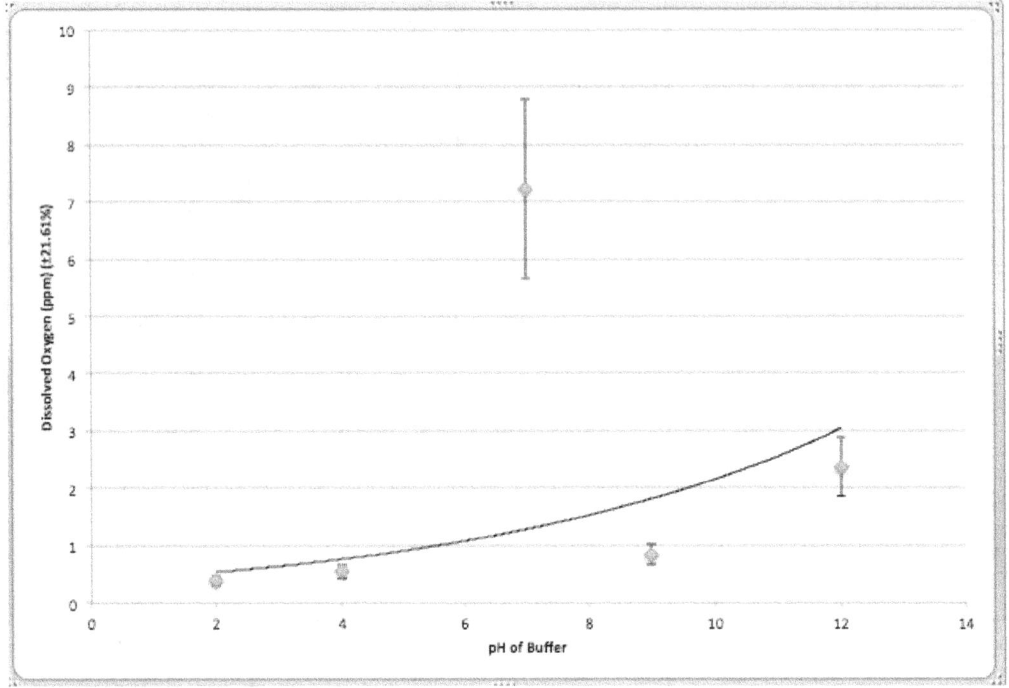

Conclusion:

The purpose of conducting this experiment and collecting this data was to look at the effects of pH on dissolved oxygen. Research concludes that as the pH increases, becoming more basic, the level of dissolved oxygen should also increase in a linear relationship.

After conducting this experiment, a conclusion has been formed. As the pH of the water increases by adding more basic buffer solution, the dissolved oxygen increases. This is evident in the graph and data collected. The exponential trendline has an exponent of 0.1727, indicating a positive change in dissolved oxygen as the pH rises. I choose the exponential trend line because of the similar labs trendlines done online.

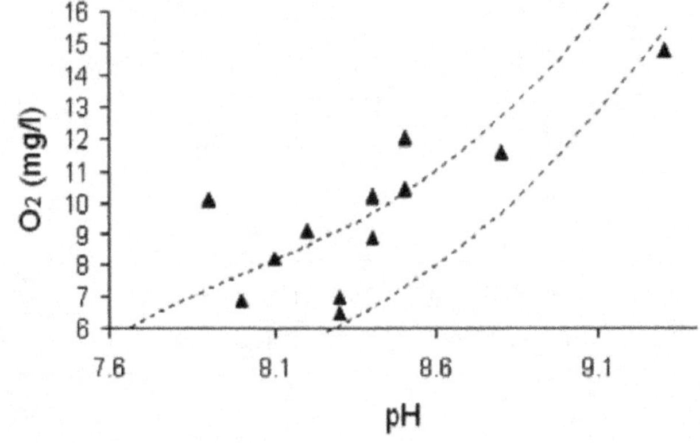

Example of Relationship of pH and Dissolved Oxygen. (25)

There is an $R^2 = 0.32391$, showing a slight disconnection between the trend line and the data. However, if we look at the data it still supports the basic conclusion that as pH increases, so does dissolved oxygen. When there was a buffer with a pH of 9 added to the water, the average amount of oxygen in parts per million in the water was 0.83 (±21.61%). Then if you look at water with a buffer with a pH of 4, it has an oxygen content of 0.54 ppm (±21.61%). The relationship in my data collected was much closer to a quadratic relationship than a linear one. There may be many reasons for this which will be thoroughly discussed in the next section.

There may be an outlier with the pH of 7 buffer, but after conducting research, the typical amount of dissolved oxygen in water with a pH of 7, is between 7-8ppm. This is considered a normal healthy water, as 7 is the normal pH for water. The rest also have levels of dissolved oxygen that would be deadly to fish, however, these are likely to never be reached in the real world.

However, my basic trend of both variables being directly related to each other is quite similar to other labs done by other professionals. A lab done by Xiangkai Wang, and Davide Baj, where they looked at dissolved oxygen and pH inside of the same water at different times. They found that as the pH dropped so did the amount of dissolved oxygen. (21)

An important aspect discussed before is how these low levels of oxygen affects aquatic life and the aquatic ecosystem. Dissolved oxygen is one of the most important components of aquatic life. A lower level of dissolved oxygen can have an increased stress on aquatic life. The recommended amount of oxygen for fish to breath through their gills is 5mg/L of dissolved oxygen. Fish start to get distressed between 2-4 mg/L. When it drops below 2mg/L, this is where death starts to occur. (7)

Solutions:

While there are already many organizations out there who are trying to tackle the issue of global warming, there are less who are monitoring the pH of water and testing it to make sure it is not hypoxic. They would also educate people who live near bodies of water to show the effects of the shifting pH of the water. A solution for this could be creating teams of volunteers whose primary jobs is to check the pH of water as well as dissolved oxygen levels using probes. They would then drop off this data at a local place that is in change. This is a good solution as it is not relatively expensive as you don't have to pay volunteers and people can do it on their own time. However, this has its own limitations as it assumes that people have their own pH/dissolved oxygen probes or if they don't have them, they must know how to use them. Another issue with just raising awareness is that it doesn't always promote change. It is impossible to force people to change their actions, so no change could come out raising awareness.

For areas where the pH is already quite low, there are many options to raise the pH of water. One of the most common ones is called liming. This is the act of adding blocks of limestone to the water to raise the pH of the water. Some advantages of this is that it is relatively cheap and limestone is non-toxic to fish and other aquatic organisms. However, there has been research that has linked these acidified lakes to more acid rain, which can have damaging effects on soil and other plants surrounding the body of water. It is also relatively hard to do in much larger bodies of water as much more limestone is required to adequately raise the pH.

Evaluation:

While I was able to obtain a conclusion that corresponds with the findings in similar labs, there are still many areas within the design of the lab that could have been improved and could have been fixed. These issues being fixed could have yielded much more accurate results that would have given a broader range of data and would have helped come to a better conclusion.

The first issue I found with my lab and process was the error of leaving the bottle with no buffer out for two days. Initially, I thought that the precipitate would need a lot longer to settle in the bottle then thought. Thus I took 300ml of water, added 2 ml of Manganese Sulfate ($MnSO_4$) and 2ml of Alkali Iodide (1 mol dm^3 of KOH and 1 mol dm^3 KI). I then left it sitting inside of a fume hood for 2 days. This is a large random error as temperature can have a large effect on the amount of dissolved oxygen in a water sample. Because the room temperature was constantly changing due to people entering and exiting and lights and air conditioning turning on and off, the dissolved oxygen was also affected by this. The amount of oxygen in typical tap water where I am from, is about 8-9ppm. My value was 7.23 ppm, which is slightly below the normal value. To fix this issue, I would have obtained all the values in one day. This makes sure that there is not a large fluctuation in temperature, so that the dissolved oxygen is not affected.

The second issue I found with my lab design was a large error in the design process. During my design is said "Stop the titration when your water becomes a pale straw color." Later in the lab, I said "Titrate the solution until it is a clear, colorless liquid." Both of these things are very subjective. How do I know that the "pale straw color" was the same color each time? This can mean that there may have been more or less Sodium Thiosulphate used, creating a random error that may have completely skewed the values. To have a definitive idea of what color is needed, there could be a certain shade of color that the solution could be compared to. This could be palate that shows the color that we desire. To be even more accurate, a colorimeter could be used to have a definitive value that is needed to stop titrating at.

The final issue with my lab is the values of pH buffers chosen. These values are likely to never be reached within the near future. I originally chose these values because of the materials readily available to me. The values of 12,9,4, and 2 are far too unrealistic, giving us a systematic error and don't give us an idea of what will actually happen with the levels of dissolved oxygen in bodies of water. To combat this issue, I would use much smaller values of pH ranging from 7-7.5 pH, as those are values that we are much more likely to see as the ocean becomes more acidic.

Extension:

To further investigate this issue, I wish to look at dissolved oxygen in actual water samples. I could look at the dissolved oxygen in relation to the time of year or water level in a body of water. There are many more possibilities to explore in terms of dissolved oxygen and many more issues to solve in terms of it. I would also love to look at how the moles of CO_2 exposed to a water sample affects pH and dissolved oxygen.

3. OCEAN SEDIMENTATION AND CARBON DIOXIDE UPTAKE BY AQUATIC PLANTS

Author: Anonymous
Moderated Mark: 29/30

<u>**Background Information & Environmental Context**</u>
Environmental Issue
Mangroves provide a range of ecosystem services including supporting biodiversity and functioning as sediment sinks (Furuwaka), however, mangrove deforestation is becoming increasingly common ("Sediments and Mangroves"). Coral reefs have a symbiotic relationship with mangroves - the mangroves trap sediment and nutrients, preventing these deposits from washing into the ocean ("Mangrove Trees & The Great Barrier Reef"). The destruction of mangroves may impact the penetration of light into surface water due to increasing deposits of terrestrial sediment into the ocean. One example of this is the Great Barrier Reef, where the loss of mangroves increases agricultural sediment, which increases the turbidity of coastal waters. Increased turbidity decreases the amount of light available for photosynthesis, influencing primary and secondary productivity of the entire reef ecosystem.

The ocean is one of the largest carbon sinks in the world, sequestering an estimated 50% of anthropogenic CO_2 emissions ("Ocean Acidification"). Increased CO_2 in the atmosphere also increases the amount of CO_2 in the water due to atmospheric-oceanic gas exchange, causing ocean acidification. CO_2 combines with seawater (H_2O) to create carbonic acid (H_2CO_3), lowering the pH of the ocean and creating a higher concentration of hydrogen (H^+) ions, posing threats to organisms sensitive to changes in acidity (Acidification Chemistry). This then reduces the number of carbonate (CO_2^{-3}) ions which are vital to shell and exoskeleton growth in many marine organisms, including coral polyps. Thus, ocean acidification is intertwined with the destruction of coral reefs.

<u>**Research Question**</u>
How does modelling changes in turbidity through decreasing available light influence the photosynthesis rates of freshwater *elodea* as indicated by pH?

Connection to Research Question and Hypothesis
This question aims to model the changes in light level caused by increasing sedimentation in the ocean. *Elodea* is a species of aquatic plant and will represent aquatic ecosystems influenced by mangrove deforestation. Based on the scientific background given for ocean acidification, it can be hypothesised that the more CO_2 in a body of water, the more acidic the water becomes. As insolation is a limiting factor to photosynthesis, it can also be assumed that the less light available for photosynthesis, less CO_2 will be used by the plan. The experiment combines these two principles in examining how sedimentation in a body of water will influence photosynthesis in aquatic producers, as measured indirectly by the pH of the water. Therefore, the hypothesis for this investigation is: as the amount of available light decreases, the pH of water containing *elodea* will decrease, becoming more acidic.

Variables
Independent and Dependent Variables

Variable Name	Measurement Range & Units	Method of Management
Independent: Turbidity: representing available light	Number of wraps of mesh around each bottle (increments 0, 3, 5, 8 and total darkness using foil).	Clearly labelling the number of wraps for each bottle to avoid confusion. Securing with elastic bands to prevent changes mid-investigation.
Dependent: pH	Measured using a pH probe (uncertainty ±0.1).	Conduct three trials for each light level increment to ensure sufficient data.

Controlled Variables

Variable Name	Method & Justification of Controlling
Water	pH will differ depending on the source of water (eg. tap, saline, distilled). As salt-water tolerant plant species are not readily available, fresh water from the lab taps will be used for all trials. This minimises potential differences in productivity between each increment.
Species of plant	Different species of plant photosynthesise at different rates. Keeping species consistent minimises discrepancies in levels of primary productivity.
Size of plant	Plants with a large mass/size typically have larger surface areas, which theoretically increases PP. By maintaining all *elodea* clippings to 10cm in length, differences in PP are minimised.
Position of bottles	Different areas of the laboratory will be exposed to different amounts of sunlight depending on the time of day; in order to control the amount of sunlight that reaches the plants, the experiment will be placed in one spot.

Materials List

Name of Resource	Quantity
Elodea aquatic plants	15 clippings, 10cm long
Graduated Nalgene plastic bottles	15
pH probe	1
Fresh water	Approx 3750ml, or enough to fill 15 bottles
Distilled water	500ml
Mesh	10cm wide strips, enough to wrap 12 bottles a total of 51 times
Aluminium foil	10cm wide strips, enough to fully cover 3 bottles
Tray	1, to hold all 15 bottles
Elastic bands	15

Masking tape	For labelling/securing mesh

Methodology

Part 1 - Setup

1. Place one *elodea* cutting into each square bottle and fill to just below the rim with fresh tap water, repeating this 15 times. Only one type of water and one species of plant must be used.
2. Cap all bottles and label trials and increments with masking tape to avoid confusion.
3. For the first increment of full light, leave three bottles unwrapped and place in the tray.
4. Cut and measure mesh strips 10 cm wide, long enough to wrap around one bottle once. Secure with masking tape and an elastic band around the rim.
5. Repeat step 4 for the remaining bottles wrapped 1 time.
6. Repeat steps 4-5 for the remaining increments of 3, 5, and 8 rounds of wrapping, measuring enough mesh to cover each bottle with the appropriate number of wraps.
7. For the final increment, cover the three bottles, including the caps, in aluminium foil, making sure no part of the bottle can be seen.
8. Place the tray containing all 15 bottles next to a window or another source of light. Ensure this position is consistent throughout the duration of the experiment (see fig. 1).

Figure 1. Final Lab Setup

Part 2 - Data Collection

1. Every two days, conduct pH readings for all trials.
2. Unscrew the cap and place the pH probe into the bottle, waiting 10 seconds for the value to stabilise before recording.
3. Resecure the mesh around the bottle and rinse the probe with distilled water between measurements.
4. Repeat steps 2-5 15 times for all bottles, for at least 3 days of data collection until sufficient data is generated.

Justification of Sampling Strategy

Changing light levels using mesh is a way of modelling increased turbidity as a result of sedimentation. As sunlight is needed for photosynthesis, if insolation is limited, the primary productivity of aquatic plants will also be limited. The number of layers of mesh for each increment was thus selected to model variations in turbidity. Three trials for each of the five increments are used to increase the validity and reliability of the experiment; the results will be measured over at least 3 separate days in order to interpret any trends in pH.

Risk Assessment and Ethical Considerations

This investigation does not deal with hazardous chemicals, however after the experiment, dispose of the aquatic plants sustainably. They can continue to be grown to minimise waste; alternatively, they can be treated as organic waste and used for composting. Remaining water from the bottles can be used to water existing plants rather than being discarded.

Data Collection

28th September

Light Level	Trial 1	Trial 2	Trial 3	Average
		pH		
0	9.10	8.90	9.10	9.03
3	9.30	9.30	9.20	9.27
5	8.10	8.00	8.30	8.13
8	8.10	8.00	8.10	8.07
foil	7.90	7.80	7.50	7.73

3rd October

Light Level	Trial 1	Trial 2	Trial 3	Average
		pH		
0	9.30	9.20	9.50	9.33
3	9.50	9.30	9.60	9.47
5	8.50	8.90	9.60	9.00
8	9.70	8.90	9.30	9.30
foil	8.10	8.70	8.10	8.30

9th October

Light Level	Trial 1	Trial 2	Trial 3	Average
		pH		
0	9.50	9.70	9.50	9.57
3	9.10	9.40	9.30	9.27
5	9.70	8.90	9.20	9.27
8	9.80	9.00	9.20	9.33
foil	9.00	8.80	8.40	8.73

Figure 2. Tables 1-3: Raw Data Over Three Days

Sample Calculation of Mean

Formula	Example
$$\bar{X} = \frac{\sum x}{n}$$	$$\bar{X} = \frac{(9.1 + 8.9 + 9.1)}{3} = 9.03$$

Qualitative Observations

- Some plants became translucent and brown, or began to disintegrate (see fig. 3). Generally, this appeared to be plants receiving less light.
- Other plants had very visible new growth, regardless of the amount of light they received (see fig. 4).

| Figure 3. Dead *Elodea* | Figure 4: New Growth on *Elodea* |

Data Processing & Analysis

The following graphs were constructed based on the averages for each day of data collection.

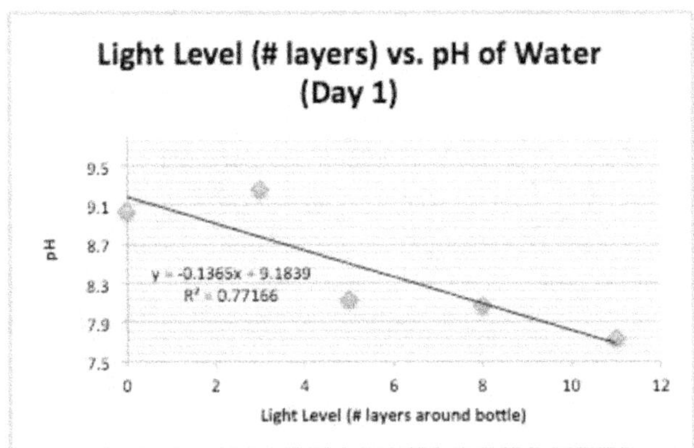

Figure 5. Day 1 Light Level vs. pH of Water

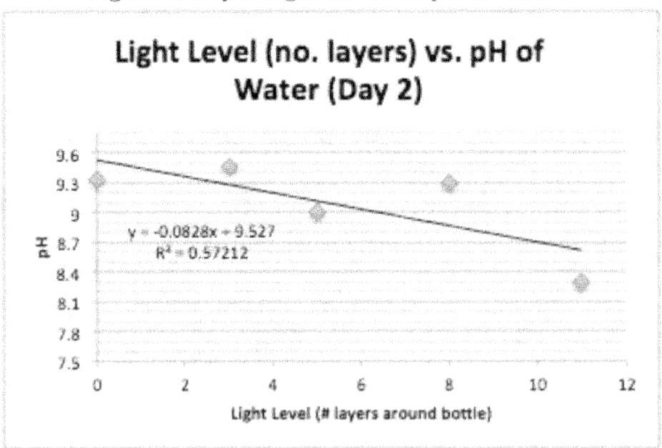

Figure 6. Day 2 Light Level vs. pH of Water

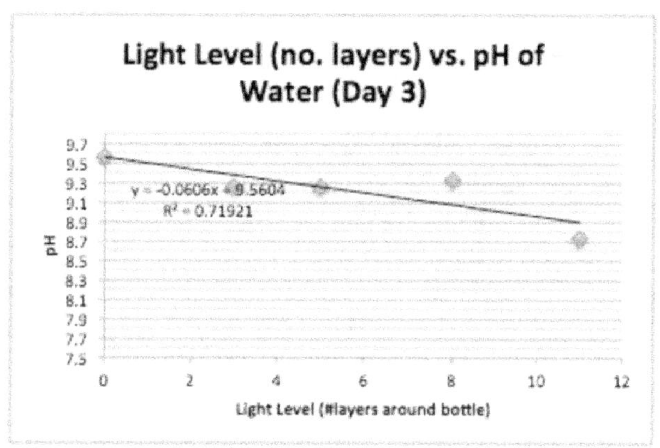

Figure 7. Day 3 Light Level vs. pH of Water

<u>Trend and Pattern in Data</u>

The r^2 value of 0.81674 for Day 1 indicates a strong correlation between light level and pH. As the amount of light available to the plant decreases, the pH of the water becomes more acidic. The bottles with full light had an average pH of 9.03, while the bottles with no light had an average pH of 7.73.

The r^2 value for Day 2 of data collection is 0.57212, indicating a weak to moderate relationship between the two variables. Still, the average pH level decreased for each increment: from 9.33 to 8.30 for the bottles with light and no light respectively. This trendline was the weakest out of all three days; notable anomalies include an increase in pH for the second increment to 9.47 before dropping suddenly to 9.00. The fluctuation in the spread of data suggests the relationship between pH and levels of light may not be as strong as originally hypothesised.

The r^2 value for the final day continues to indicate a moderately strong correlation between light levels and pH, at 0.71921. The pH begins at 9.57 and decreases to 8.73. The collective change in the pH over the 5 increments for all three days can clearly be observed to decrease. As available light is reduced, so the pH of the water becomes more acidic.

An observation of the averages for each increment was that despite the pH decreasing for each increment on any given day, the overall pH of the bottles continued to increase. For example, the bottles with no light began with a pH of 7.73, which then increased to 8.73 by the end of the experiment. These changes are represented in the following graph.

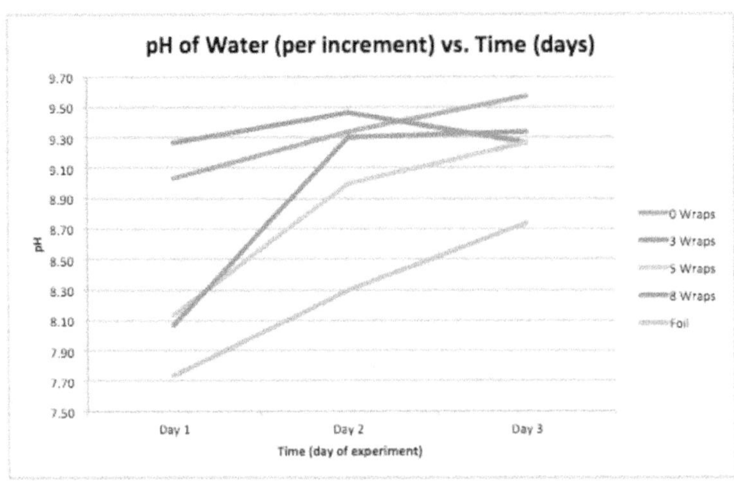

Figure 8. Changes in pH of Water vs. Time

Nearly all increments (except for the bottles with 3 wraps of mesh) are shown with an increase in pH over time, indicating that as the experiment progressed, the pH became increasingly alkaline.

Conclusion

The above analysis suggests that the pH of water decreases as the amount of light decreases, aligning with the original hypothesis. Though not an original part of the hypothesis, it is interesting to note the progression of pH over time towards alkalinity, rather than acidity. Despite the fact that the data supports the original hypothesis, there are certainly limitations to the method that may reduce the reliability of the data.

Evaluation

In Context of Environmental Issue

Based on the results of this experiment, changing light levels (representing increasing ocean sedimentation due to mangrove deforestation) does have an impact on aquatic pH. This supports the idea that mangrove ecosystem services benefit surrounding ecosystems like coral reefs. However, when considering the use of aquatic producers to reduce the impacts of ocean acidification, the results of this experiment do not align with this theory. The conclusion demonstrates an increase in alkalinity of the water as light was reduced, rather than an increase in acidity. Therefore, a decrease in ocean acidity cannot be reliably associated with the presence of aquatic producers.

Evaluation of Method

Weaknesses of the Experiment

The biggest limitation of this experiment is the uncertainty of the pH probe, used to measure the dependent variable. During the experiment, it was noted that two different pH probes produced readings with a difference of up to 2 (eg. one probe would read '7.5' and the other would read '9.5'), thus potentially impacting the results of this particular experiment.

Unawareness of how to care for aquatic plants like *elodea* may have impacted the rate of photosynthesis. It was noted in the qualitative observations that *elodea* became translucent or brown as the experiment progressed, indicating a deterioration to the health of the plant. This would have an impact on the ability of the plants to conduct photosynthesis, and thus the amount of carbon dioxide they were able to absorb, which would potentially impact the pH of the water.

<u>Strengths of the Experiment</u>
The use of mesh to control the amount of available light for each plant was appropriate, as was the number of wraps chosen for each increment. The gradual changes in light through the use of mesh were able to support the changes in pH seen from the experiment. Furthermore, the control variables were maintained relatively successfully. The water and species of plant used was the same for all trials, helping to control the reliability of results to some extent.

Improvements and Further Areas of Research
The bottles were placed upright during the experiment, providing a low surface area for light to penetrate (see fig. 1 - Final Setup). The close alignment of the bottles may have lead to different trials receiving differing amounts of light. One improvement to address this issue and thus improve the reliability of the experiment would be to use an artificial light source (a grow light) instead of natural light. This would allow a more consistent, higher concentration of light to reach the bottles, which may yield more distinct differences in pH. The ± 0.1 uncertainty for the digital pH probe used to collect data could be addressed through the use of a different pH probe connected to a digital program such as LoggerPro. This would allow changes in pH to be tracked over time, and more accurate conclusions regarding the influence of light levels on pH to be drawn.

<u>Applications</u>
Application to Environmental Context
The photosynthesis and primary productivity of producers removes roughly 25% of CO_2 from the atmosphere ("Plants Absorb More CO_2"). As one form of solution for rebalancing oceanic pH, aquatic primary producers can be introduced to increase rates of photosynthesis. Phytoplankton blooms and photosynthesis can be encouraged through iron fertilisation, increasing the amount of carbon dioxide absorbed from the atmosphere ("Ocean Acidification: Geoengineering"). Organic carbon absorbed by blooms at the surface is sequestered into mid-ocean waters when decomposers and zooplankton consume the phytoplankton ("Fertilising the Ocean With Iron"), effectively preventing the carbon from continuing to acidify vulnerable surface ocean ecosystems. In this investigation, the *elodea* plants represent the phytoplankton as both are producers, helping to remove CO_2 from an aquatic environment. The results of the experiment indicate that water increases in alkalinity as light levels are reduced, meaning in coastal reef areas affected severely by both ocean acidification *and* sedimentation, aquatic producers could form part of the solution to reducing the acidity of the water.

Evaluation of Effectiveness
However, based on the minimal reliability of the data from this experiment, it is unclear whether the use of aquatic producers like phytoplankton will have long term success in mitigating ocean acidification. Additionally, in real life, there is insufficient evidence to suggest that as light levels are reduced, the alkalinity of a body of water increases. Small iron fertilisation experiments have been conducted (12 since 1993), however the long-term impacts of increased phytoplankton blooms on aquatic food webs have not been fully assessed. Out of the 12 experiments, only 3 demonstrated carbon sequestration by producers ("Fertilising the Ocean With Iron"). Some detrimental impacts of using aquatic producers as a method of carbon sequestering include the depletion of essential nitrates and phosphates in the water, which could alter the available nutrients for fish and other secondary consumers.

When considering mangrove deforestation and increases in sediment deposits, this solution may not be as viable, as insolation is a key factor limiting the primary productivity of aquatic plants, which has indirect impacts on the amount of CO_2 these producers are able to remove from the oceans. Furthermore, this solution does not directly address the roots of the environmental issue - the deforestation of mangroves which results in increased sedimentation, as well as the increase in anthropogenic carbon dioxide emissions which increase the acidity of ocean water, limiting its effectiveness.

Works Cited

Furukawa, Keita, and Eric Wolanski. "Sedimentation in Mangrove Forests." *Mangroves and Salt Marshes*, vol. 1, no. 1, Mar. 1996, pp. 3-10, doi:10.1023/A:1025973426404. Abstract.

"Mangrove Trees & the Great Barrier Reef." *Canisius Ambassadors for Conservation*, Institute for the Study of Human-Animal Relations, www.conservenature.org/learn_about_wildlife/great_barrier_reef/mangrove_trees.htm.

"Ocean Acidification." *National Geographic*, Apr. 2017, www.nationalgeographic.com/environment/oceans/critical-issues-ocean-acidification/.

The Ocean Portal Team. "Ocean Acidification." Edited by Jennifer Bennett. *Ocean Portal*, Smithsonian Institution, ocean.si.edu/ocean-life/invertebrates/ocean-acidification.

"Plants Absorb More CO2 than We Thought, but ...." *The Conversation*, Oct. 2014, theconversation.com/plants-absorb-more-co2-than-we-thought-but-32945.

Powell, Hugh. "Fertilizing the Ocean with Iron." *Oceanus Magazine*, vol. 46, no. 1, Jan. 2008, www.whoi.edu/oceanus/feature/fertilizing-the-ocean-with-iron.

Swales, Andrew, editor. "Sediments and Mangroves." *National Institute of Water and Atmospheric Research*, Nov. 2012, www.niwa.co.nz/freshwater-and-estuaries/research-projects/sediments-and-mangroves.

4. AN INVESTIGATION INTO INVERTEBRATE AND TREE DIVERSITY OF A NATIVE FOREST AND A PINE PLANTATION OF ABERDEENSHIRE, SCOTLAND

Author: Robert Craig

I. Identifying the Context

<u>Research Question</u>

Does species diversity of family of trees and terrestrial invertebrates vary significantly between the native forest of Drum Estate and the pine plantation of Crathes Estate (of Aberdeenshire), as measured using Simpson's diversity index?

<u>Environmental Issue: The decreasing tree and invertebrate diversity due to deforestation and the increasing number of forest plantations</u>

Living in a wildlife rich country, I was interested to investigate species diversity. This interest was further developed when we completed an ESS field trip, where we investigated freshwater species diversity and its abundance in polluted and unpolluted waters. This field work inspired me to research the tree and terrestrial invertebrate diversity found in an old, native forest ("Drum Castle, Garden & Estate"), and a man-made pine plantation of Aberdeenshire ("Crathes Castle, Garden & Estate").

Today, only one third of global forest area is composed of native forests ("Forests"). Global trends suggest that the rate of deforestation of native forest has been intensely increasing since 1990, while "the areas of forest plantations are increasing" ("Forest"). 'Deforestation' is a much discussed topic. While they are economic benefits to it, such as answering the human demands for cattle, crops, wood, and more living space, the replacement of native habitats by plantations also results in a "substantial loss of biodiversity" ("Loss of Biodiversity (including Genetic Diversity)"). Habitat destruction and fragmentation impacts the less resilient species of ecosystems, decreasing their population numbers (Fitzherbert, B. & al.). Deforestation and plantations also contribute to problems such as global warming and unsustainable practices ("Palm Oil and Tropical Deforestation").

<u>Connection of Environmental Issue to Research Question</u>

This global issue is connected to the research question as deforestation has "greatly reduced Scotland's 'native' tree cover" (Hall), thus altering the "biodiversity and ecosystems in Scotland" ("Climate Change"). Simpson's diversity index have been used in past studies to observe species and tree diversity of Scotland.

To investigate this local environmental issue, I observed the species diversity and individual abundance of families of invertebrates (found in leaf litter) and trees (found above quadrats) of two different forest types. I then completed Simpson's diversity index to test whether the diversity significantly varies between the "native, ancient forest of Drum Estate" ("Drum Castle, Garden & Estate") and the man-made pine plantation of Crathes Estate, dating from the mid -206' century ("Crathes Castle, Garden & Estate").

II. Planning

<u>Justification with Hypothesis</u>

I expect to see a higher tree and invertebrate diversity in the native forest than in the pine plantation. This is because the pine plantation is more recent, making its species diversity lower. The plantation of only pine trees has also reduced the tree diversity.

<u>Justifying Sample Strategy</u>

To collect data for this experiment, I used a sampling strategy using $0.25m^2$ quadrats in pre-determined areas at pre-determined coordinates. This strategy enabled for data to be quickly and efficiently obtained, and assured that I observed a representative and statistically accurate sample of species and tree diversity in both forests.

<u>Variables Identification and Explanation</u>

Independent: Type of forest (native woods or pine plantation)

Dependent: Insect and tree species diversity

Controlled: Approximately similar total area investigated; same species identification key; same measurement tools; same time allowed at each invertebrate identification (12 minutes); same observer for all records; recorded on the same weekend (similar weather)

<u>Equipment List</u>

- Thermometer ($C°$)
- Digital pH metre
- pH 9.2, 4, and 7 buffer solutions
- Pipettes
- Insect and tree family identification keys
- One $0.25m^2$ quadrat
- Plastic spoon
- Magnifying glass
- Empty data tables and pen
- Camera
- Timer
- Petri dish
- 10 clean plastic bags to collect soil samples
- 2 white trays

<u>Risk Assessment and Ethical and Environmental Considerations</u>

When calculating the soil pH and handling acidic and basic solutions, use gloves and wear a lab coat and protective goggles. Avoid all direct contact with the solutions. If contact is made, rinse immediately and inform your teacher. If glass materials breaks, inform your teacher and pick up the pieces using gloves.

When observing the insect and tree species, make sure no organism is harmed, harassed, injured, or killed. Use a spoon (with no sharp edges) to pick up organisms. When placing individuals in the petri dish or tray, handle them with care. Always release the organisms and the leaf litter in the same area as they were found. When collecting soil sample, make sure no organisms are enclosed in the bag.

<u>Methodology</u>

A Establishing the sampling area and coordinates

1. Find a native forest and a recent plantation in the same region of a country (eg: Aberdeenshire).

2. Using the Free Map Tools website ("Area Calculator Using Maps"), select the "Area Calculator Using Maps" option and calculate the sampled area (km^2), making sure it is similar in size for both forests. Use the same website every time.

3. Using the website Grid Reference Finder ("UK Grid Reference Finder"), select 5 coordinates, spreading them out throughout the previously selected area (all points must be within the selected area). The coordinates must be chosen ahead of time in order to limit bias. I later realized that I should have used a random number generator to generate these coordinates — this weakness will be explored in the discussion. Use the same website every time.

4. Print out a map of your selected area and coordinates to refer to them during the field experimentation (see Figure 1).

Area Output

252531.575 m²
0.253 km²

<u>Figure 1</u>: *Left and above*: The selected sampling area of Crathes' pine plantation in km^2 (using the website Free Map Tools). Drum's selected sample area measured 0.247 km^2.

Bellow: Pre -selected coordinates within Crathes' pine plantation selected area (using the website Grid Reference Finder). The same method was used to obtain the coordinates for Drums' native forest.

Grid Reference	X (Eastings)	Y (Northings)	Latitude	Longitude
NO 72883 97063	372883	797063	57.063805	-2.448776
NO 72927 97192	372927	797192	57.064970	-2.448059
NO 73175 97105	373175	797105	57.064198	-2.443957
NO 73064 97010	373064	797010	57.063341	-2.4457830
NO 73134 97253	373134	797253	57.065529	-2.4446565

1. Go to the selected area of one of the forest (native). Using a grid reference app (Irving) on your mobile phone, go to your selected coordinates. Use the same app every time.

2. Place the centre of the quadrat on the floor at the coordinate. If there are logs or large rocks in the way, slightly adjust the location, keeping it as close as possible to the original one.

3. Record your abiotic factors. Measure the air temperature 5 cm above the quadrat using the same thermometer ($C°$). Take a soil sample (2 small handfuls) from the centre of the quadrat and place it in an enclosed plastic bag. The pH will be tested later. Avoid collecting rocks, twigs, leaves, or organisms.

4. Record into the data table the number of individuals of each tree family whose branches pass over the quadrat (when looking up from the centre of the quadrat; see Figure 4).

5. Record any relevant qualitative data such as weather, soil moisture, or changes in location.

6. Start your timer, setting it to 12 minutes.

   a) Collect leaf litter from inside the quadrat and place it in a white tray.

   b) Using a species identification key (Bebbington), record all the families of the invertebrates found and the number of individuals from each families. Make sure not to count one organism twice by placing the recorded ones in a different tray.

      i.   Use a magnifying glass or petri dish for accurate identification.

   c) Stop all invertebrate identification once the timer rings.

7. Replace the observed soil and invertebrates into the quadrat area, making sure nothing is harmed in the process.

8. Repeat steps 1-7 for all 5 coordinates site, making sure to test the entire leaf litter of each quadrat, then go to the second forest area (plantation) and repeat steps 1-7 again. Use the same material every time.

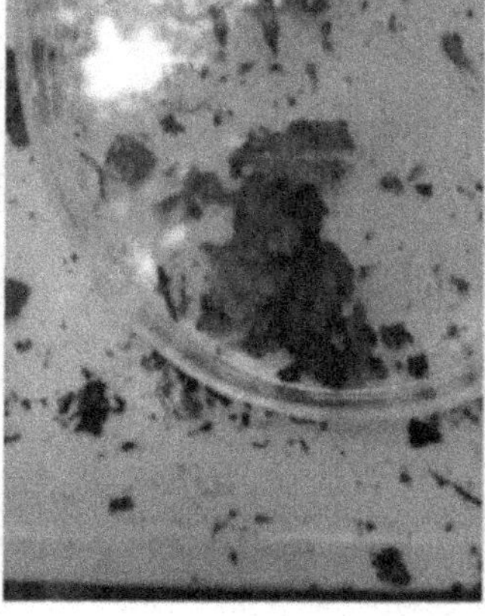

Figure 2: A quadrat placed down at specific coordinates in the old native woods of Drum.

Figure 3: An invertebrate placed in a petri dish for better identification.

Figure 4: Branches of trees which are found above a specific record quadrat; this method is used to record tree diversity (see step 4).

C. Testing the pH of the Sampled Soil

1. This procedure takes place in a lab. Use the same electronic pH probe for each test.
2. Connect the pH probe to a computer and open a Logger Pro recording sheet.
3. To test the pH's calibration, place it in pH 7, pH 9.2, and pH 4 buffer solutions to make sure it shows the right measurements on the Logger Pro graph.
4. Once the pH probe has been correctly calibrated, place the soil samples in different, clean beakers, each appropriately labelled with their site number (1-5) and forest types.
5. Add distilled water to the beakers until all the soil is covered. Use a clean stirring rod to stir each soil and water solutions. Let them sit for 5 min (this insures that any rocks or large debris sink to the bottom).
6. Press "Collect" on a new Logger Pro graph collection document, and carefully place the waterproof part of pH probe in the beaker. Stir the pH probe in the solution for 92 seconds (to reach a stable average). Once finished, press "Stop" and save the document with the appropriate name (site number and forest area).
7. Repeat steps 5-6 until pH for all of the soil samples has been recorded.
8. Copy the data for the 92 seconds into an excel sheet, making sure to label the correct quadrats and locations. Using Excel's "Average" option, obtain the mean average for the pH of each site and copy it into the raw data tables.

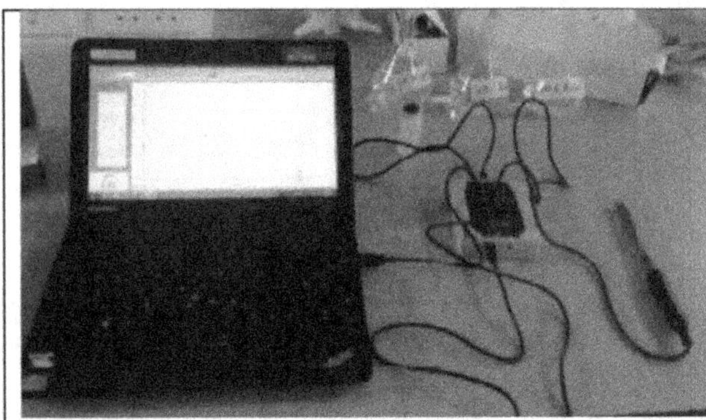	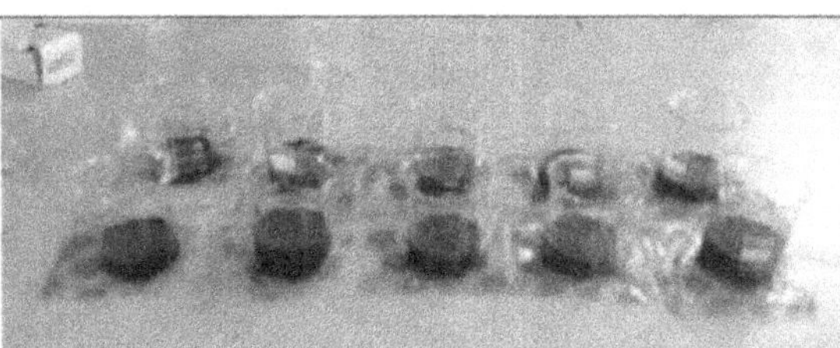
Figure 5: A computer plugged to the pH probe using LoggerPro.	*Figure 6:* The sample soil from all 10 quadrat of the two locations mixed with water before being tested for their pH.

III. Results, Analysis, and Conclusion

A Drum Estate's Native Wood Raw Data

<u>Table 1:</u> Number of individuals found for each invertebrate and tree species, as well as abiotic measurements as sampled in 5 quadrats in the old, native woods of Drum Estate.

	Quadrat 1 NJ 79411 00598 (+/- 5 metres)	Quadrat 2 NJ 79252 00738 (+/- 5 metres)	Quadrat 3 NJ 79166 00876 (+/- 5 metres)	Quradrat 4 NJ 79358 00914 (+/- 5 metres)	Quadrat 5 NJ 79401 00639 (+/- 5 metres)
Family of Invertebrate (includes their larvae stages)	Total Number of Individuals (+/- 1)	Total Number of Individuals (+/- 1)	Total Number of Individuals (+/- 1)	Total Number of Individuals (+/- 1)	Total Number of Individuals (+/- 1)
Snails					
Slugs		1			
Earthworms	2	1			1
Beetles	2	1	1	1	
True Bugs	2	1		3	3
True Flies	1	4		2	3
Bees/Wasps					
Spiders	4	5	3	5	7
Woodlice					
Centipedes					
Ants					
Butterflies/Moths					
Earwig	3	1	1	1	2
Millipedes					
Mite	2				3
Harvestmen	1	1			
Springtail		6	2	2	2
Family of Tree	Total Number of Individuals (+/- 1)	Total Number of Individuals (+/- 1)	Total Number of Individuals (+/-1)	Total Number of Individuals (+/- 1)	Total Number of Individuals (+/- 1)
Elm	4		1		
Oak			2	1	
Lime		1			1
Birch	1				9
Pine					
Abiotic Factors					
Air Temperature (C°; +/- 0.1 C°)	14	14	15	14	14
Soil pH (rounded to 3SF; +/- 0.01)	4.26	4.33	3.90	3.93	4.01
Observations	Overall, all data was collected in soft ground surfaces - in fallen leaves, under rocks and logs, no more than 2cm under the surface of the soil; no quadrats moved from original coordinates				

Table 2:

Number of individuals found for each invertebrate and tree species, as well as abiotic measurements as sampled in 5 quadrats in the pine plantation of Crathes Estate.

	Quadrat 1 NO 73175 97105 (+/- 5 metres)	**Quadrat 2** NO 73064 97010 (+/- 5 metres)	**Quadrat 3** NO 72883 97063 (+/- 5 metres)	**Quradrat 4** NO 72927 97192 (+/- 5 metres)	**Quadrat 5** NO 73134 97253 (+/- 5 metres)
Family of Invertebrate (includes their larvae stages)	Total Number of Individuals (+/- 1)	Total Number of Individuals (+/- 1)	Total Number of Individuals (+/- 1)	Total Number of Individuals (+/- 1)	Total Number of Individuals (+/- 1)
Snails					
Slugs					
Earthworms					
Beetles					
True Bugs	1			1	1
True Flies	3	2	2		2
Bees/Wasps			1 (wasp)		
Spiders	4	4	2	6	4
Woodlice				1	
Centipedes					
Ants					
Butterflies/Moths					
Earwig					
Millipedes	1				
Mite					
Harvestmen					
Springtail	2	2		1	
Family of Tree	Total Number of Individuals (+/- 1)	Total Number of Individuals (+/- 1)	Total Number of Individuals (+/-1)	Total Number of Individuals (+/- 1)	Total Number of Individuals (+/- 1)
Elm					
Oak					
Lime					
Birch					
Pine	5	3	2	4	2
Abiotic Factors					
Air Temperature (C°; +/- 0.1 C°)	15	15	15	16	18
Soil pH (rounded to 3SF; +/- 0.01)	4.02	4.27	4.14	4.60	4.20
Observations	-sampled in an area with a mix of needles, pine twigs, and decaying leaves	-sampled in area with mainly needles and pine twigs	-sampled in a more compact soil, harder to dig into and more humid; sample not taken more than 2cm below leaf litter surface	-sampled in a more compact soil, with a thinner layer of pine needles; one small rock with moss on top found inside quadrat	-sampled in decaying leaves, clovers, thinner layer of pine needles

C. Justification of Values of Error

The values of error for the thermometer and the pH probe correspond to their smallest units of measure. The number of individual invertebrates and trees have an uncertainty of +1- 1 as I may not have seen every organism. The values of error for the coordinates correspond to the ones given by the app (Irving).

D. Processed Data

Table 3:Total number of individuals for each invertebrate and tree species, as well as mean pH for both forests

Native Forest of Drum Estate						
Total number of invertebrates recorded	Quadrat 1	Quadrat 2	Quadrat 3	Quadrat 4	Quadrat 5	Total (all quadrats)
	17	21	7	14	7	66
Total number of trees recorded	5	1	3	1	10	20
Total mean temperature (for all quadrats; 3SF) and Standard Deviation	14.2 °C 0.447 SD					
Total Mean pH (for all quadrats; 3SF) and Standard Deviation	4.09 0.197 SD					

Pine Plantation of Crathes Estate						
Total number of invertebrates recorded	Quadrat 1	Quadrat 2	Quadrat 3	Quadrat 4	Quadrat 5	Total (all quadrats)
	11	8	5	9	7	40
Total number of trees recorded	4	3	2	4	1	14
Total mean temperature (for all quadrats; 3SF) and Standard Deviation	15.8 "C 1.30 SD					
Total Mean pH (for all quadrats; 3SF) and Standard Deviation	4.25 0.218 SD					

E. *Calculating Simpson's Diversity Index*

Calculation I: Simpson's diversity index for invertebrates in the old native woods of Drum Estate (note that all species of 1 are not included as 1-1 = 0)

Native Forest of Drum Estate	Quadrat 1	Quadrat 2	Quadrat 3	Quadrat 4	Quadrat 5
Simpson's Diversity Index: $d = \dfrac{N(N-1)}{\sum n(n-1)}$ Note: diversity rounded to 3SF	$N = 17 \cdot 16 = 272$ $n = 2*1 + 2*1 + 2*1 + 4*3 + 3*2 + 2*1 = 26$ $d = \dfrac{272}{26} = 10.5$	$d = \dfrac{420}{62} = 6.77$	$d = \dfrac{42}{8} = 5.25$	$d = \dfrac{182}{30} = 6.07$	$d = \dfrac{420}{64} = 6.56$
Average Diversity of Invertebrates (calculated using Excel) & **Standard Deviation**	7.03 1.49 SD				

Calculation 2: Simpson's diversity index for invertebrates in the pine plantation of Crathes Estate (note that all species of I are not included as 1-1 = 0)

Pine Plantation of Crathes Estate	Quadrat 1	Quadrat 2	Quadrat 3	Quadrat 4	Quadrat 5
Simpson's Diversity Index: $$d = \frac{N(N-1)}{\sum n(n-1)}$$	$d = \frac{110}{20} = 5.50$	$d = \frac{56}{16} = 3.50$	$d = \frac{20}{4} = 5.00$	$d = \frac{72}{30} = 2.40$	$d = \frac{42}{40} = 3.00$
Average Diversity of Invertebrates (calculated using Excel) & Standard Deviation	3.88 1.30 SD				

Calculation 3: Simpson's diversity index for tree diversity in the native woods of Drum Estate

Native Forest of Drum Estate	Quadrat 1	Quadrat 2	Quadrat 3	Quadrat 4	Quadrat 5
Simpson's Diversity Index: $$d = \frac{N(N-1)}{\sum n(n-1)}$$	$d = \frac{20}{12} = 1.67$	$N = 1*0 = 0$ $n = 1*0 = 0$ $d = undefined$	$d = \frac{6}{2} = 3.00$	$N = 1*0 = 0$ $n = 1*0 = 0$ $d = undefined$	$d = \frac{90}{72} = 1.25$
Average Diversity of Invertebrates (calculated using Excel) & Standard Deviation	1.97 1.84 SD				

<u>Calculation 4</u> Simpson's diversity index for tree diversity in the pine plantation of Crathes Estate

Pine Plantation of Crathes Estate	Quadrat 1	Quadrat 2	Quadrat 3	Quadrat 4	Quadrat 5
Simpson's Diversity Index: $d = \dfrac{N(N-1)}{\sum n(n-1)}$	$d = \dfrac{20}{20} = 1.00$	$d = \dfrac{3}{3} = 1.00$	$d = \dfrac{2}{2} = 1.0$	$d = \dfrac{12}{12} = 1.00$	$d = \dfrac{2}{2} = 1.0$
Average Diversity of Invertebrates (calculated using Excel) & **Standard Deviation**	1.00 0 SD				

Graph 1: Processed data comparing the total number of individuals of invertebrates of each family for both types of forest

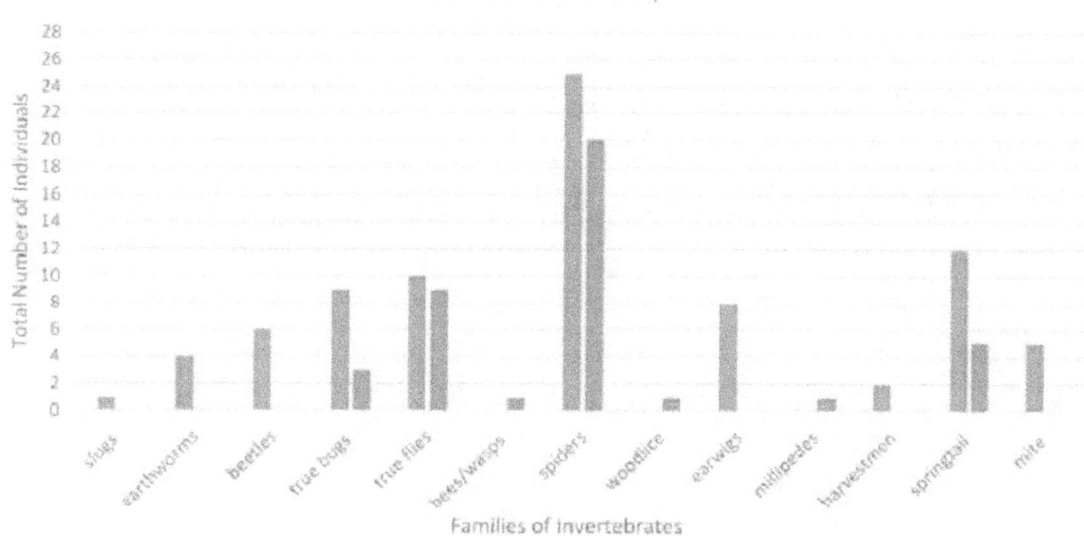

Graph 2:Processed data comparing the total number of individuals of trees of each family for both types of forest

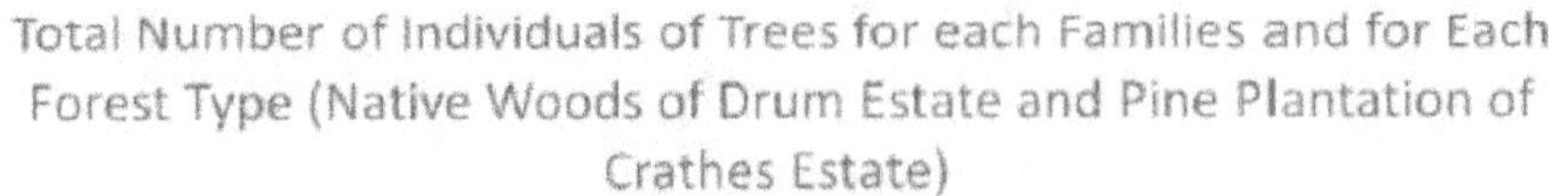

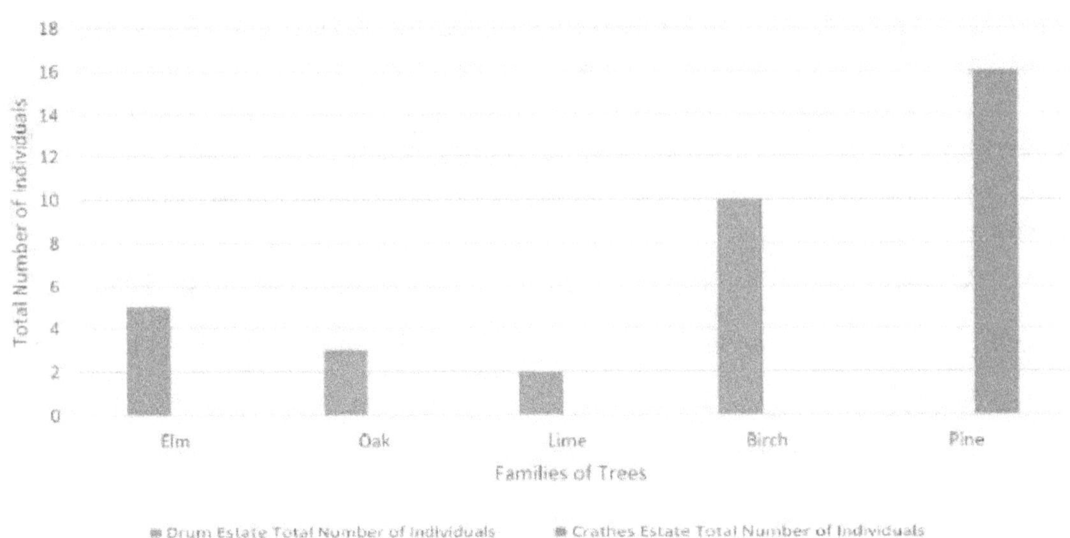

Graph 3: Average invertebrate diversity in both forest types as measured using Simpson's Diversity Index.

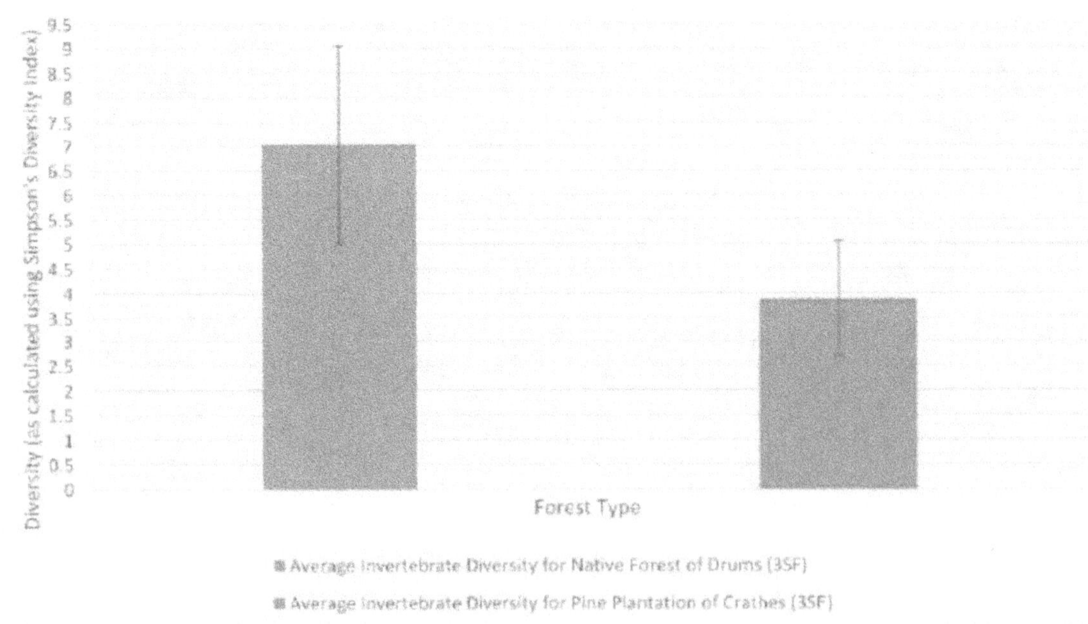

Graph 4: Average tree diversity in both forest types as measured using Simpson's Diversity Index.

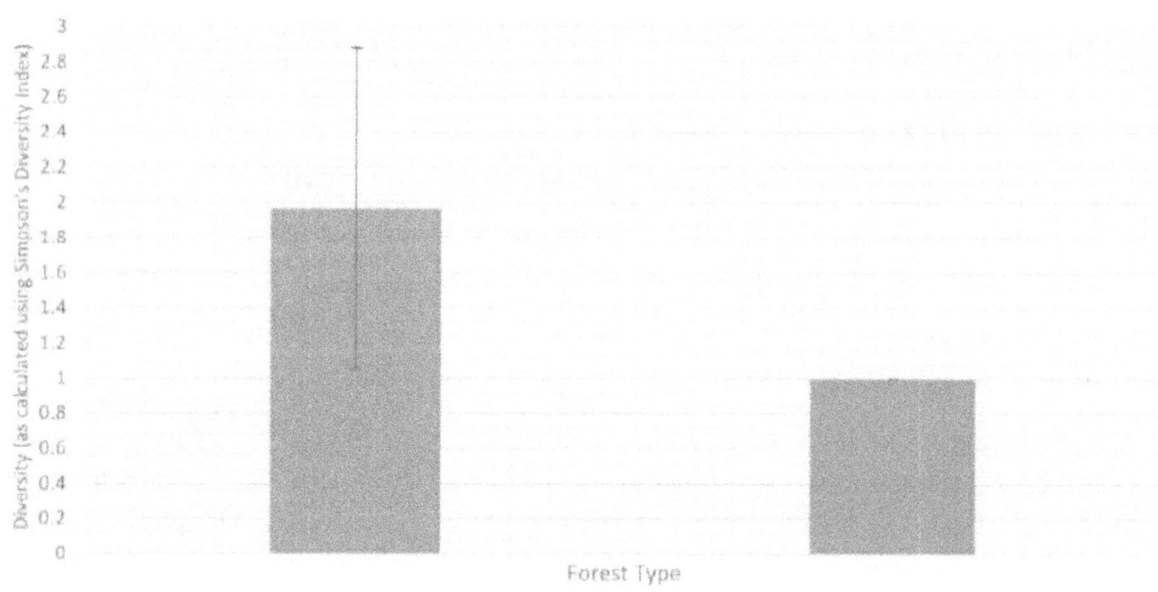

F. Calculating the t-value

Table 4: T -test value for the invertebrate and tree diversity of both forest types

	Invertebrate Diversity	**Tree Diversity**
T -test value (calculated using Excel; rounded to 3SF)	0.0104	0.0449
Percentage chance that the null hypothesis is true	1.04%	4.49%

To obtain the t-test value, I used the t -test formula on Excel by plugging in the average diversity of each quadrat for the invertebrates and the trees, using a 2 tail. The values obtained are smaller than 5%, which means that the null hypothesis can be rejected. The null hypothesis would be that there is no significant difference between the tree and invertebrate diversity of the two forests. Despite an overlap between the error bars of the invertebrate diversity graph, the t –test value implies that the two means are significantly different.

G. Conclusion

Graphs 1 and 2 demonstrate that there is a greater variety and abundance of invertebrate and tree species in the native forest. Some species, such as millipedes and woodlice, are found in the pine plantation and not the native woods. This indicates that some species have adapted to particular habitat, influenced by factors like pH (which is slightly more acidic in the pine plantation).

Graphs 3 and 4 show that there is an overall greater diversity of invertebrate and trees in the native forest of Drums (7.03 and 1.97 versus 3.88 and 1). These findings suggest that this habitat is healthier and composed of more complex food webs, making it more resilient. The diversity of 1 for trees of the Crathes plantation is due to the fact that pine trees were the only type of trees present, and thus, despite their abundance, the lack of variety resulted in a low diversity.

V. Discussion and Evaluation

A. Evaluation of Conclusion in Context of Environmental Issue

This conclusion shows that native habitats are more diverse and thus more resilient to changes. Therefore, the issue of deforestation could be resolved by conserving native habitats over plantations, as "plantations tend to support a very limited range of wildlife" ("Deforestation"). While there are economic benefits to plantations, these unsustainable conservation practices result in an eventual extinction of endemic species found in some native forests (eg: Amazon rainforest). Deforestation and replantation also contribute to other environmental issues like climate change and habitat fragmentation ("Palm Oil and Tropical Deforestation").

B. Discussion of Strengths. Weaknesses, and Limitations of Method

Further areas of research include recording species diversity at more quadrat stations and in an intermediate habitat.

Strengths of this investigation include the effectiveness of the procedure, which was quick to complete in both habitats, produced sufficient data, and which can be easily repeated in different habitats. The 'limitations land weaknesses of this investigation were that there was some uncertainty in the recordings of species diversity, as some organisms may not have been counted or wrongly identified. Additionally, a random number generator should have been used to produce coordinates in the forest areas. A modification of this investigation would thus be to use a random number generator on Excel to produce coordinates in a chosen area.

VI. Applications

A. Potential Solution to Environmental Issue

Creating 'conservation areas to protect native forests is a sustainable solution to biodiversity loss and deforestation. For example, the successful Ngorongoro Crater conservation site in Tanzania now provides ideal conditions "for the survival and preservation of a large diversity of species" ("Welcome To Ngorongoro Crater Destination Guide"). This solution has already been applied to the native forest of Drum Estate, but more conservation areas around Aberdeenshire could be created in order to preserve other native habitats.

B.	Evaluation of the Solution

A strength of this solution is that it would provide locals with job and ecotourism, thus contributing to a country's growth all while conserving its species. A weakness is that it is hard to determine where these conservation areas should be (as species may migrate) and is expensive to maintain. It would also result in the collapse of other industries that are dependent on the resources found in these habitats (eg: wood industries).

VII. Bibliography

"Area Calculator Using Maps." *Free Map Tools.* N.p., n.d. Web. 25 Oct. 2016.
https://www.freemaptools.com/area-calculator.htm.

Bebbington, Anne, John Bebbington, and Steve Tilling. *The Woodland Name Trail.* N.d. A key to the invertebrates of soil and leaf litter. Strafford Park, Telford.

"Climate Change." *Biodiversity Scotland.* N.p., n.d. Web. 06 Nov. 2016.
http://www.biodiversityscotland.gov.uk/biodiversity/pressures/climate-changet

"Crathes Castle, Garden & Estate." *The National Trust For Scotland.* N.p., n.d. Web. 06 Nov. 2016. http://www.nts.org.u1c/Property/Crathes-Castle-Garden-and-Estate/

"Deforestation." *Trees for Life.* N.p., n.d. Web. 29 Sept. 2016.
http://treesforlife.org.uk/forest/human-impacts/deforestation/.

"Drum Castle, Garden & Estate." *The National Trust For Scotland.* N.p., n.d. Web. 06 Nov. 2016. http://www.nts.org.uk/Property/Drum-Castle-Garden-and-Estate/.

Fitzherbert, Emily B., Matthew J. Struebig, Alexandra Morel, Finn Danielsen, Carsten A. Bruhl, Paul F. Donald, and Ben Phalan. "How Will Oil Palm Expansion Affect Biodiversity?" *Review.* Creative Educational Press, n.d. Web. 22 Sept. 2016. <Downloads/fitzherbert_2008_tree_how%20will%20oil%20palm%20expansion%20affect%20biodiversity.pdf>.

"Forests": 2. *How Much Forest Is There on the Planet and at What Rate Is It Disappearing?* Green Facts, n.d. Web. 29 Sept. 2016. <http://www.greenfacts.org/en/forests/1-2/2-extent-deforestation.htm>.

Hall, Jeanette. "Forest and Woodland: Woodland Cover." *Scottish Natural Heritage.* N.p., n.d. Web. 18 Sept. 2016. http://wwvv.snh.org.uk/publications/on-line/advisorynotes/138/138.htm.

Irving, Mike. "Grid Ref UK and Ireland App for IPhone, Android, Windows." Mike Irving. *Apps* Np, nd Web. 25 Oct 2016 <http//www.mike-irving.co.uk/portfolio/mobile-apps/grid-ref-uk-and-ireland/>.

"Local Nature Conservation Sites." (n.d.): 4. Web. 25 Oct. 2016. <https://www.aberdeenshire.gov.uk/media/11110/localdevelopmentplan2016-proposedplan-supplementaryguidance5a-Incssitesindex_000.pdf>.

"Loss of Biodiversity (including Genetic Diversity)." *Rainforest Conservation Fund.* N.p., 02
Mar. 2010. Web. 06 Nov. 2016. <http://www.rainforestconservation.org/rainforest-
primer/3-rainforests-in-peril-deforestation/f-consequences-of-deforestation/3-loss-of-
biodiversity-including-genetic-diversity/>.

"Natural Heritage Trends Forest and Woodland: Woodland Cover." *Scottish Natural Heritage.*
N.p., n.d. Web. 15 Nov. 2016. http://www.snh.org.uk/publications/on-
line/advisorynotes/138/138.htm.

"Palm Oil and Tropical Deforestation." *Union of Concerned Scientists.* N.p., n.d. Web. 25 Oct.
2016. http://www.ucsusa.org/global_warming/solutions/stop-deforestation/palm-oil-and-
forests.html#.WA-WCOgrKhc.

"UK Grid Reference Finder." *Grid Reference Finder.* N.p., n.d. Web. 25 Oct. 2016.
http://www.gridreferencefinder.com/

5. WHAT EFFECT DOES THE PERCENTAGE CHANGE IN THE (HDI) HAVE ON PERCENTAGE CHANGE IN (EF) PER CAPITA, FROM 2000 TO 2016, OF THE 28 (EU) NATIONS

Author: Adam Zhao
Moderated Mark: 27/30

With the depletion of natural capital at a rate faster than the rate replenished, the ecological overshoot over the carrying capacity is an issue of global concern. According to the Global Footprint Network, it is estimated that humanity uses the equivalent of 1.6 planets needed for resources yearly. This is expected to increase to two planets by 2050. The deficit between biocapacity and planetary biodiversity has also decreased by 30% since 1970, where ecological footprint (EF), defined as the measurement of "is the area of land and water required to sustainably provide all resources at the rate at which they are being consumed by a given population" ("Global Biodiversity has Declined") and measured in global hectares (gha), is the culprit.

The global problem of a high EF is primarily caused by carbon emissions produced from fossil fuels, which constitutes 50% of total EF (see figure 1) (Kahn, 2017). It is also the fastest growing cause, where its rate has increased 100 to 200 times faster since the ice age (Kahn, 2017). This is correlated with the rate of urbanization doubling in the past 25 years (Kahn, 2017). Economic development has led to the exploitation of fishing grounds, cropland, built-up land, forest products, and grazing land that introduce greenhouse gases into the atmosphere, destroy habitats, and decrease biodiversity. These consequences which make up the EF, amplified by the fact that the agricultural sector now occupies 33% of land surface and 75% of freshwater resources ("UN Report: Nature's Dangerous Decline 'Unprecedented'; Species Extinction Rates 'Accelerating'"), make it is no surprise that EF per capita has increased globally. Increased demand for an increasing population, with 7.7 billion people worldwide (Lenzen, et. al., 2013) without replenishment leads to a growing scarcity of resources. Per capita consumption is also accountable with increased wealth and technology and its high energy usage, powered mostly by fossil fuels.

Figure 1: World Ecological Footprint by Land Type from 1961-2016

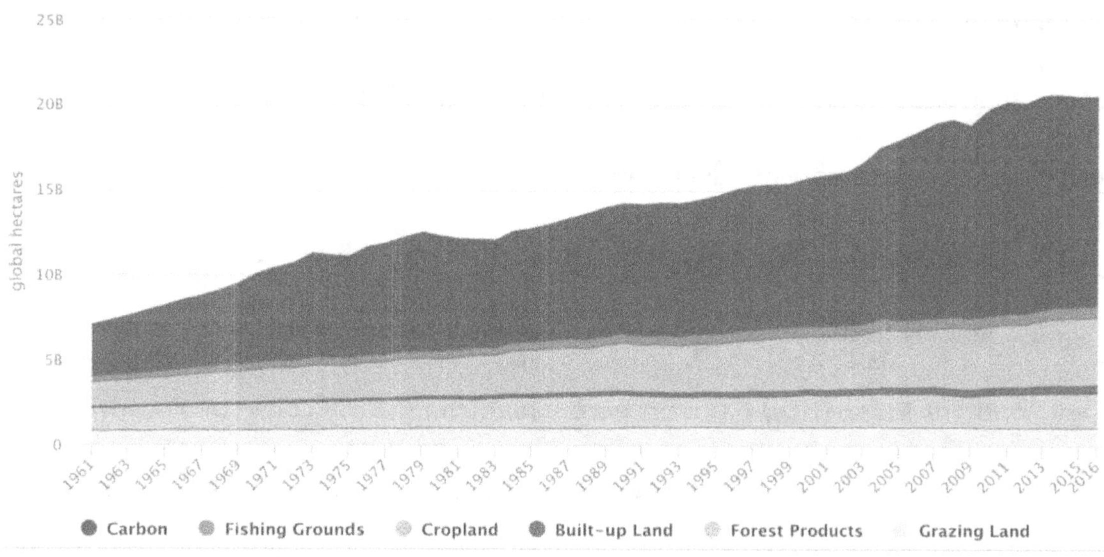

(Kahn, 2017)

In the instance where these limited resources are shared by a given community, where principles of the Tragedy of the Commons are present, there is a multitude of consequences worldwide. The carbon footprint causes the enhanced greenhouse effect and its impacts such as atmosphere contamination, runoff into the hydrosphere from ice melting and thermal expansion causing extreme weather events leading to the displacement of settlements and many fatalities. This, in addition to land clearing in the biosphere, compounds to the extinction of endemic flora and fauna that offer ecosystem goods, like food and medicine, costing up to $125 trillion annually (Mclendon, 2018). In addition, the carbon footprint leads to positive feedback. For instance, in the melting permafrost in tundra biomes caused by greenhouse gases releases methane, contributing to more greenhouse gases in the atmosphere.

Sustainable development, defined by the UN as "meeting the needs of the present without compromising the ability of future generations to meet their own needs" ("Sustainable Development Agenda") is the necessary mindset, where everyone, including future generations, can thrive in one planet. However, the debate lies on whether this can still be achieved with a high standard of living, thus leading to the research question, **"What effect does the percentage change in the Human Development Index (HDI) have on percentage change in Ecological Footprint (EF) per capita, from 2000 to 2016, of the 28 European Union (EU) nations?"** This links to the global issue of high EF as composite factors of the independent variable of life expectancy, education, and per capita income can either raise environmental awareness, the overconsumption of goods by increased ability to become consumers, or the opposite of such. This investigation aims to find out if there is a wealth or development level at which EF is most minimized.

II. Planning

<u>i) Method</u>

Table 1: Variables

Type of Variable	Description
Independent	HDI Percentage Change between 2000 to 2016
Dependent	EF per Capita Percentage Change between 2000 to 2016
Controls	- The starting and ending years for determining percentage changed are placed at 2000 to 2016. - The database for research will only be obtained from the Global Footprint Network (GFN) and Human Development Reports. (HDR) situated within it. - The geographical coverage of this study is kept at the 28 EU nations only. - The same laptop will be used with an internet connection and with Excel software and database compatibility for GFN

<u>Procedure:</u>

1. Access GFN's Open Data Platform at data.footprintnetwork.org.

2. Open Explore Data Header.

3. Select Sustainable Development Category.

4. Filter to HDI and EF Variables.

5. Filter Geographical Selection to Europe.

6. Adjust Time Slider to year 2000.

7. Click Download Data.

8. Repeat steps 6 and 7 for year 2016.

9. In an Excel File, remove all non-EU nations.

10. Organize data into the following Table 2.

Table 2: Sample data table template

	EF per capita in 2000 in gha (to 2 dps)	HDI in 2000 (to 2 dps)	EF per capita in 2016 in gha (to 2 dps)	HDI in 2016 (to 2 dps)
Austria				
Belgium				
Bulgaria				

<u>ii) Justification of Sampling Strategy</u>

The EU is selected as the sampled countries since its countries are of a high wealth level and thus a good sample for showing if wealthy countries are able to change consumption and waste habits. It is one of the few transnational governance groups that operate in the same region with the same regulations, thus acting as a control.

The HDI indicator was chosen as a holistic measurement for life expectancy, education, and per capita income, rather than just economic development. EF was justified for its consideration of both land and water area needed to be consumed at a sustainable rate.

The time frame chosen between 2000 to 2016 is an additional measure to see progress in sustainability made after the Millenium Development Goals in correlation with HDI as an indicator of development for society in three measures.

The adequate number of countries, set at 28, allows for application of a t-test, which determines how strong the relationship is between two variables with a monotonic function. It is also a large sample out of the Organisation for Economic Co-operation and Development countries, which involve high HDI levels ("OECD").

<u>iii) Safety and Ethical Issues</u>

Though all research will be carried out online, the Pomodoro study technique shall be applied to minimize stress, back pain, and fatigue, as well as eye strain from radiation, whilst maximizing efficiency through interval breaks. Correct posture and the UV-mediation Flux software would also be used to help for the back pain and eye strain factors respectively.

Ethical implications would include engaging in secondary research sources that respect autonomy and transparency. The Global Footprint Network is a credible source as evidenced by its collaboration with WWF, Earth Day Network, and various federal governments.

III. Data Presentation, Analysis, and Conclusion

Table 3: Raw data table showing EF per capita and HDI for 2000 and 2016 for the 28 EU member nations

	HDI in 2000 (to 2 dps)	EF per capita in 2000 in gha (to 2 dps)	HDI in 2016 (to 2 dps)	EF per capita in 2016 in gha (to 2 dps)
Austria	0.84	5.68	0.91	6.03
Belgium	0.87	7.69	0.92	6.25
Bulgaria	0.71	3.04	0.81	3.45
Croatia	0.75	3.20	0.83	3.94
Cyprus	0.80	5.33	0.87	3.75
Czechia	0.80	5.58	0.88	5.59
Denmark	0.86	8.83	0.93	6.80
Estonia	0.78	6.02	0.87	7.06
Finland	0.86	6.23	0.92	6.26
France	0.85	5.54	0.90	4.45
Germany	0.87	5.51	0.93	4.84
Greece	0.80	6.40	0.87	4.27
Hungary	0.77	3.52	0.83	3.61
Ireland	0.86	6.36	0.93	5.12
Italy	0.83	5.60	0.88	4.44
Latvia	0.73	4.27	0.84	6.36
Lithuania	0.76	3.64	0.86	5.57
Luxembourg	0.86	14.94	0.90	12.91
Malta	0.78	6.46	0.88	5.79

Netherlands	0.88	6.30	0.93	4.83
Poland	0.79	4.26	0.86	4.43
Portugal	0.79	4.83	0.85	4.10
Romania	0.71	2.39	0.81	3.09
Slovakia	0.76	3.73	0.85	4.21
Slovenia	0.82	4.80	0.89	5.13
Spain	0.82	5.56	0.89	4.04
Sweden	0.90	6.43	0.93	6.46
United Kingdom	0.87	5.73	0.92	4.37

Table 4: Processed data table showing EF per capita and HDI percentage change for the 28 EU member nations between 2000 and 2016

	HDI % change 2000 and 2016 (to 2 dps)	EF % change 2000 and 2016 (to 2 dps)
Austria	8.33	6.16
Belgium	5.75	-18.73
Bulgaria	14.08	13.49
Croatia	10.67	23.13
Cyprus	8.75	-29.64
Czechia	10.00	0.18
Denmark	8.14	-22.99
Estonia	11.54	17.28
Finland	6.98	0.48
France	5.88	-19.68
Germany	6.90	-12.16
Greece	8.75	-33.28

Hungary	7.79	2.56
Ireland	8.14	-19.50
Italy	6.02	-20.71
Latvia	15.07	48.95
Lithuania	13.16	53.02
Luxembourg	4.65	-13.59
Malta	12.82	-10.31
Netherlands	5.68	-23.31
Poland	8.86	3.99
Portugal	7.59	-15.1
Romania	14.08	29.29
Slovakia	11.84	12.81
Slovenia	8.54	6.8
Spain	8.54	-27.34
Sweden	3.33	0.4
United Kingdom	5.75	-23.7
Mean	**8.84**	**-2.5**
Standard Deviation	**3.07**	**22.5**

Sample Calculations

<u>Mean ($\underline{x}$):</u>

$$\underline{x} = \frac{\sum_{i=1}^{n} x_i}{n}$$

Substitute for HDI% Change Mean

$$\underline{x} = \frac{8.33+5.75+14.08...+5.75}{28} = 8.84$$

Standard Deviation (s):

$$s = \sqrt{\frac{\sum_{i=1}^{n} (x_i - \mu)^2}{n}}$$

Substitute for HDI% Change Standard Deviation

$$s = \sqrt{\frac{(8.33-8.84)^2+(5.75-8.84)^2+(14.08-8.84)^2...+(5.75-8.84)^2}{28}} = 3.07$$

Percentage Change:

$$\frac{N - I}{I} \times 100\%$$

where N indicates New Value and I indicates Initial Value

eg. Substituting for Austria Data Point for HDI:

$$\frac{0.91 - 0.84}{0.84} \times 100\% = 8.33\%$$

A two tailed *t*-test is conducted due to the testing of statistical significance.

Identify null hypothesis: There is no difference between the percentage change of HDI and the percentage of EF from 2000 to 2016

Identify alternative hypothesis: There is a statistical difference between the percentage change of HDI and the percentage of EF from 2000 to 2016

$$t = \frac{\bar{x}_1 - \bar{x}_2}{\sqrt{\frac{s_1^2}{n_1} + \frac{s_2^2}{n_2}}}$$

Where subscripts 1 and 2 refers to HDI and EF respectively, $\underline{x}$ refers to means of data sets, s refers to standard deviation, and n refers to the sample size.

Substitute values:

$$= \frac{2.55 + 8.84}{\sqrt{\frac{22.57^2}{28} + \frac{3.07^2}{28}}} = 2.65$$

Comparative critical value in t-table, with 26 degrees of freedom from df = n-2

t-score is 1.706 under 0.05 significance

2.65 > 1.706, therefore the null hypothesis is rejected, and there is a statistical significance in the relationship between HDI and EF percentage change over 2000-2016

Figure 2: Scatter plot graph showing the relationship between EF per capita and HDI percentage change over time for 28 EU member nations

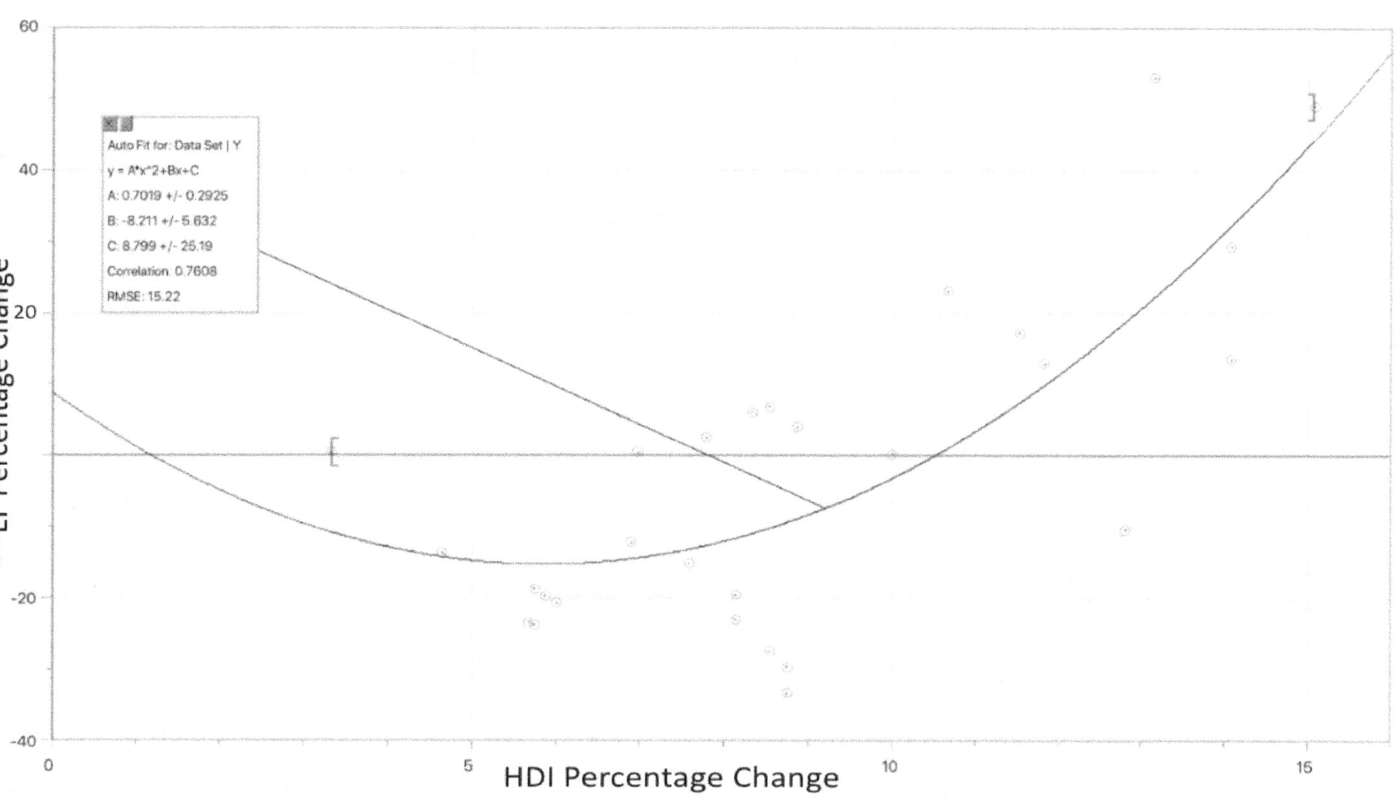

The GFN states that a minimum HDI score of 0.7 is considered high development while a minimum EF score of 1.7 gha per capita is considered replenishable ("Global Footprint Network"). When looking at Figure 2, all the EU member nations satisfy the HDI score but are well above the sustainable EF score, to varying extents. Figure 1 denotes the positive quadratic regression present in the trend of the relationship between HDI percentage change and EF percentage change. Hence, the extreme ends of the spectrum, where little development and high development changes are associated with high rates of environmental degradation, and vice versa. The correlation coefficient (r) is 0.76, indicating a strong relationship between these two variables.

Most countries follow this general trend, as seen in Sweden at the extreme left with an HDI change of 3.3% and EF change of 0.5%, and Lithuania at the extreme right with 13.2% and 53.0% for the same respective values. However, there are anomalies, with midpoint values exceeding that of Sweden's. Mainly, Austria has an HDI change of 8.3% and EF change of 6.2%, when other countries around that HDI change, located at the minimum point of the graph, such as Spain, has values of 8.5% and -27.3% respectively. This may be due to high priorities set upon green urban infrastructure, conservation sites, and general public perception for ecocentric habits (Lenzen, et. al., 2013).

All of these countries are situated in the later stages of the Demographic Transition Model, i.e. Stage 4 and 5, where urban facilities are completed. It may involve the acquisition of green technologies that have been centered around technocentric, costly innovations. One might associate such advancements to a lesser EF, but rather, when considering the scale of facilities, where less than 20% of infrastructure is allocated towards renewables, the excess usage of energy will lead to the increase of EF ("Global Footprint Network"). It has also been found that the world's richest countries in terms of GNI per capita like the United Kingdom and Denmark consume on average 10 times as many materials as the poorest ("Global Footprint Network") while Greece, with minimal economic growth, though still rich, would rather allocate its resources to business endeavors. Therefore, one can make the conclusion that moderately-developed nations on HDI yields lowest EF; once above a certain threshold of wealth and development, and the affording of green technology with energy limits, they have the optimal level of EF.

IV. Discussion and Evaluation

This study links to the global issue of environmental degradation in correlation to wealth as it shows that countries are moving in the right direction ecologically if they are developed at a certain threshold, yet should have economic limits so excess unsustainable development does not occur. Priority should be set that the tipping point that moves lower-income countries to green economies.

The biggest strength of this investigation was the reliability of the data collected from renowned multinational organizations. All data came from one website, i.e., Global Footprint Network, had a control of coming from the same years, therefore, leading to an absolutely valid conclusion, without bias. Having 28 data points also allowed for a strong correlation.

The most major weakness is how the calculation of percentage change does not factor what occurs between the start and finish year, and economic recessions or environmental disasters like oil spills are not shown on the data. Also, as these organizations carry these out throughout a large geographical area, the measurements of HDI and EF are hard to measure and may have inaccuracies. As a composite indicator, HDI may also have flaws in its measurements, like when one of such, mean years of education, does not consider the quality of education.

Further research would need to be conducted rather than modifications to this existing one due to the nature of weaknesses. In addressing the multifaceted, difficult to measure indicators, another independent variable would be looked at. GNI per capita would be analyzed instead of development as a whole, while carbon footprint as the dependent variable would replace the broader ecological footprint, both now disregarding percentage change.

V. Application

The solution proposed is to prioritize the allocation of resources and natural environment to aid development, where each individual nation should identify which is their largest EF contributor. Since this is usually carbon, they should create leapfrog schemes from fossil fuels to green technology. This is seen in recent measures to the United Nations Framework Convention on Climate Change, specifically the Paris Agreement, drafted in 2015, which recognizes a bottom-up approach ("Paris Agreement"). This instills consensus building while the nationally determined targets involves the interconnectedness between socioeconomic sectors and environmental management. If more priority is explicitly set in the articles about quantitative findings about how the environment drives developmental progress, its impact will be strengthened. Emphasis on getting funding for the Green Fund which leads into National Adaptation Programmes of Action can help higher income nations to support lower income ones to achieve green technology.

This solution is beneficial in the way that it is straight to the point by means of the collaboration between member nations. It also does not necessitate the need for new development strategies. As an anthropocentric measure, it is unlike ecocentric ones which need a lot of time, or technocentric ones, which may be costly. Since existing measures of international cooperation are already in place such as the Paris Agreement, as stated in the aforementioned, this solution develops on an existing framework, rendering processes of the discussions feasible and efficient.

The main weakness of this solution is how rich countries may want their money to be allocated into their own personal development rather than aiding other nations. Countries where carbon emissions are not as big of an issue as the other causes of EF, such as fishing grounds, would also have this strategy be ineffective, though this solution does address the largest root cause. Overall, the cornucopians who were previously worried about economic gains getting in the way of environmental rehabilitation measures should consider the holistic development benefits brought upon by reducing EF.

VI. Works Cited

Camdessus, Camille. "How Europe Is Faring on Renewable Energy Targets." Phys.org,

Phys.org, 31 Mar. 2019, phys.org/news/2019-03-europe-faring-renewable-energy.html.

"Global Biodiversity Has Declined." WWF,

wwf.panda.org/knowledge_hub/all_publications/living_planet_report_timeline/lpr_2012/h

ealth_of_our_planet/.

"Global Footprint Network." Global Footprint Network, www.footprintnetwork.org/.

Kahn, Brian. "Carbon Dioxide Is Rising at Record Rates." Climate Central, 10 Mar. 2017,

www.climatecentral.org/news/carbon-dioxide-record-rates-21242.

Lenzen, Manfred, and Shauna A Murray. "The Ecological Footprint – Issues and Trends ."

Integrated Sustainability Analysis, 2013,

www.isa.org.usyd.edu.au/publications/documents/Ecological_Footprint_Issues_and_Trends.pdf.

McLendon, Russell. "11 Startling Stats about Earth's Disappearing Wildlife." MNN, Mother

Nature Network, 30 Oct. 2018,

www.mnn.com/earth-matters/wilderness-resources/blogs/11-startling-stats-about-earths-

disappearing-wildlife.

"OECD." OECD, www.oecd.org/about/members-and-partners/.

"Paris Agreement." United Nations, 2015,

unfccc.int/sites/default/files/english_paris_agreement.pdf.

"The Effects of Climate Change." NASA, NASA, 9 July 2019, climate.nasa.gov/effects/.

"The Human Footprint." WWF, World Wildlife Fund,

www.worldwildlife.org/threats/the-human-footprint.

"The Sustainable Development Agenda - United Nations Sustainable Development." United

Nations, United Nations, www.un.org/sustainabledevelopment/development-agenda/.

"UN Report: Nature's Dangerous Decline 'Unprecedented'; Species Extinction Rates

'Accelerating' - United Nations Sustainable Development." United Nations, United

Nations,www.un.org/sustainabledevelopment/blog/2019/05/nature-decline-unprecedented-

report/.

6. WHAT IMPACT DOES THE IMPLEMENTATION OF "HOY NO CIRCULA" HAVE ON LEVELS OF CO, SO2, PM_{10}, NO_2 AND OZONE IN MEXICO CITY

Author: Julia Garcia

Research Question:

What impact does the implementation of "Hoy No Circula" have on levels of CO, SO2. PM_{10}, NO_2 and Ozone in Mexico City over time?

Hypothesis-

If the Implementation of "Hoy No Cireula" is effective then, pollution levels for CO, SO2, PM_{10}, NO2 and Ozone will decrease in Mexico City over time because there will be less % of cars circulating.

Mexico City is one of the largest metropolis in the world, and has struggled with air pollution for the past decades. According to Mexico City's air pollution agency, 37.4 % of air pollution is caused by cars The fact that the city is located in a valley, trapping the pollution inside makes the problem worst.

In order to fight the ever increasing pollution, Mexico City's government implemented "Hoy No Circula" in 1989 (Mario Molina) which translates to "Today [your car] does not circulate" (abbreviated to HNC). This is a law stating that all cars older than 6 years of age could not be driven once per week. In 2011, to further combat air pollution, the government decided to discard the 6 year rule, opting for a harsher one: Cars that exceed a I% CO volume could not be driven once a week.

This lab focuses on illustrating the relation between "HNC" and five air pollutants (CO, NO2, Ozone, PM_{10}, and SO2) to see if the legislation was successful. The pollutants were chosen because these are the ones produced mostly by cars. These 5 variables help tighten the scope. So there can be a closer correlation between air quality and cars.

By finding the trend of pollutants before and after the law, one can tell if this has had any effect on mitigating pollutants. The lab will also seek to see if there is a correlation between the number of cars that were kept off the streets because of "HNC" and pollution levels.

It is important that we measure the success of this law because it can help the lives of millions of Mexicans. These pollutant levels directly affect the health of people living in the city. According to Mario Molina Center. "long-term exposure shows an increase between 6 and 17% in all-cause mortality for the entire population". Measuring the success of the program matters because, if it's not working, the government must look for other solutions in order to limit the health hazard. If this project is successful, then it can be replicated in cities that suffer from similar cases.

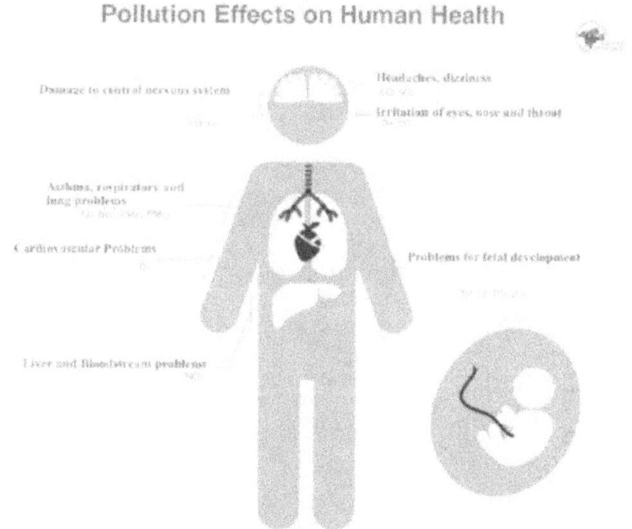

Illustration 1: Effects of Pollution on Human Health (Translated from http://www.aire.cdmx.gob.mx, 2014 air quality report)

Planning

The time range (1986-2015) is to see a relationship of contaminants before and after "HNC", as well as looking for a relationship between the number of cars that were not driven that year and the pollution level, and the average percent change.

Procedure:

1) Obtain the levels of pollution for CO, SO_2, PM_{10}, NO2 and Ozone from 1986-2015. (aire.cdmx.gob)

2) Calculate percent difference yearly from each pollutant. (equation, explanation below)

3) Find the average of percent difference for each pollutant before and after HNC (explained below)

4) Obtain the number of cars that circulated yearly in Mexico City (INEGI)

5) Obtain the number of cars that did not circulate once a year in Mexico City due to "HNC" (INEGI.)

6) Find the average of cars that did not circulate per year by dividing the number of cars that did not circulate in that year by the total number of cars.

7) Make a chart with the data from steps 4-6. Make sure that the data used begins in 1986 up to 2015. (one column per step)

8) Create a bar graph from data found in step 3. Compare percent difference average before and after the law.

9) Create a scatter plot comparing levels of one pollutant with the percentage number of cars that did not circulate. X axis the the % of cars for that given year, and the y axis will be the pollution level for that same year. (data found in steps 1 and 6)

10) Make the same scatter plot for the 4 other pollutants

11) Add straight line of best fit (if possible) for the graphs made in steps 9, 10, to see if there is a correlation between the percentage of cars and the pollution levels.

*No risk or ethical considerations apply- Justifications for procedure listed below.

Variables Table

	Named Variable	Units	Equipment or Procedure for measurement/ control
Independent	HNC implementation- Number of cars that do not circulate because of this.	Years relative to number of cars- percentage	Published data Derive percentage of cars that did not circulate (given below)
Dependent	Pollutant level of CO, NO2, Ozone, PM_{10}, SO2	CO = PPM NO2, Ozone, SO2 = PPB PM10= yg m3	Published data.

Control Variable Table

What is being controlled?	Why is it being controlled?	How is it being controlled?
Type of Pollutant	To focus on pollutants caused by cars, as "HNC" attempts to mitigate these. If this included other pollutants, they would have no relationship to the law.	By only using data from the five pollutants.
Location- Mexico City	To focus only on the pollution levels of Mexico City. As HNC was made in Mexico City, the pollutant levels used must be from this area only to stay relevant.	Only using published data from Mexico City
Years (1986-2015)	To compare the pollution levels before and after the program was implemented, see if there is any correlation between the initiation of this program and the decrease of pollutants.	Only use published data from these years.

Results, analysis and conclusion

Table 1: Number of Cars and Pollutant Levels through the years

*No uncertainties were given

Year	Number of Cars	Number of Cars that do not circulate once a week	Carbon Monoxide (CO) Levels Concentration (PPM) Particles per million	Nitrogen Dioxide (NO2) Concentration (PPB) Particles per billion	Ozone Concentration (PPB)	Particulate Matter 10 (PM$_{10}$) Concentration yg m^{-3}	Sulphur dioxide (SO2) Concentration (PPB)
1986	1,519,411	0	4.0	44.1	45.0	-	43.0
1987	1,549,203	0	3.87	41.3	34.9	-	46.3
1988	1,677,570	0	3.92	39.4	40.1	-	49.8
1989	1,711,583	273,853	4.18	40.6	44.3	175.0	51.6
1990	1,977,554	321,409	4.74	41.8	42.5	138.3	51.4
1991	1,919,557	309,929	6.29	40.4	44.0	103.6	52.8
1992	2,257,243	384,746	6.05	43.0	65.2	82.6	61.4
1993	2,566,043	462,898	4.31	45.1	54.7	96.2	32.1
1994	2,016,038	443,635	3.48	39.8	44.9	95.7	19.8
1995	2,132,325	469,322	3.01	32.7	43.2	92.5	18.7
1996	2,067,206	454,790	2.43	38.2	42.8	84.5	18.3
1997	2,100,283	457,862	2.41	35.6	39.1	79.3	17.7
1998	2,541,957	566,856	2.26	30.8	38.6	83.1	15.6
1999	2,631,169	589,328	2.20	28.4	38.2	76.6	14.8
2000	2,511,543	560,074	2.18	30.1	36.4	74.6	17.5
2001	2,407,362	536,842	2.08	27.3	36.7	58.9	18.5
2002	2,321,702	513,416	1.94	29.1	34.5	59.2	12.4
2003	2,260,123	180,810	1.88	33.2	31.6	64.3	11.7
2004	2,556,032	580,219	1.73	33.1	29.1	58.6	13.1
2005	2,696,220	609,354	1.46	32.9	27.4	53.6	11.3
2006	3,079,690	692,945	1.41	31.7	28.1	51.7	10.6
2007	3,423,719	777,184	1.36	31.2	27.6	46.3	7.9
2008	3,922,587	898,231	1.13	30.9	27.2	49.6	7.4
2009	4,120,535	935,361	1.01	29.4	25.7	53.4	6.5
2010	4,166,756	941,687	0.75	27.5	18.9	52.8	5.4
2011	4,396,912	1,125,610	0.83	27.4	24.3	62.7	3.8
2012	4,615,276	1,186,131	0.79	27.1	24.6	50.9	6.1
2013	4,960,630	1,264,960	0.74	26.7	23.7	50.2	6.7
2014	5,012,750	1,283,265	0.72	25.8	22.4	45.8	5.6
2015	5,259,553	1,341,192	0.78	25.5	23.1	35.3	4.5

Data Processing

Percent Difference, Justification:

Equation: (new-old)/old x 100

Example: Percent Difference 1990-1991 CO

1990= 4.74 (old)

1991= 6.29 (new)

(6.29-4,74)- 1.55

1.55/4.74 = 0.327

0.327 x 100 = 32.7

"old" is the previous year, "new" is the next year.

Meaning: 32.7 means that from 1990 to 1991, CO increased by 32%

Importance: Illustrates the yearly change in percentage, which simplifies all the numbers. Helps see how much a pollutant increased or decreased yearly.

Tells us if the number of cars or pollutants increased compared to the previous year.

Table 2- Percent Difference per year for number of cars and levels of pollution.

Year	Carbon Monoxide (CO) Levels Concentration (PPM, %)	Nitrogen Dioxide (NO2) Concentration (PPB, %)	Ozone Concentration (PPB, %)	Particulate Matter 10 (PM$_{10}$) Concentration - yg m^3 (%)	Sulphur dioxide (SO2) Concentration (PPB, %)
1986-1987	-3.25	-6.35	-22.4	NA	7.68
1987-1988	1.29	-4.60	14.9	NA	7.56
1988-1989	6.63	3.05	10.48	NA	3.62
1989-1990	13.4	2.96	-4.06	-20.9	0.39
1990-1991	32.7	-3.35	3.53	-25.1	2.72
1991-1992	-3.82	6.44	48.2	-20.3	16.3
1992-1993	-28.8	4.88	-16.1	16.46	-47.7
1993-1994	-19.3	-11.8	-17.9	-0.52	-38.3
1994-1995	-13.5	-17.9	-3.79	-3.34	-5.56
1995-1996	-19.3	16.8	-0.93	-8.65	-2.14
1996-1997	-0.82	-6.81	-8.64	-6.15	-3.28
1997-1998	-6.22	-13.5	-1.28	4.79	-11.9
1998-1999	-2.66	-7.79	-1.04	-7.82	-5.13
1999-2000	-0.91	5.99	-4.71	-2.61	18.2
2000-2001	-4.59	-9.30	0.82	-21.0	5.71
2001-2002	-6.73	6.59	-5.99	0.51	-33.0
2002-2003	-3.09	14.1	-8.41	8.61	-5.65
2003-2004	-7.98	-0.30	-7.91	-8.86	11.9
2004-2005	-15.6	-0.60	-5.84	-8.53	-13.74
2005-2006	-3.42	-3.65	2.55	-3.55	-6.19

Year	Carbon Monoxide	Nitrogen Dioxide	Ozone	Particulate Matter 10	Sulphur dioxide
2006-2007	-3.55	-1.58	-1.78	-10.45	-25.5
2007-2008	-16.9	-0.96	-1.45	7.21	-6.33
2008-2009	-10.6	-4.86	-5.51	7.66	-12.2
2009-2010	-25.7	-6.46	-26.5	-1.12	-16.9
2010-2011	10.7	-0.36	28.6	18.8	7.41
2011-2012	-4.82	-1.09	1.23	-18.7	5.17
2012-2013	-6.33	-1.48	-3.66	-1.38	9.84
2013-2014	-2.70	-3.37	-5.49	-8.76	-16.4
2014-2015	7.64	-1.18	3.13	-22.9	-19.6

Average Percent Difference, Justification:

Add the percent differences from a category and divide it by the number of years the percentage differences were obtained from.

Example for SO2 before "HNC".

7.68+5.56+3.62+0.39= 17.25

17.25/ 4= 4.31

The average will be divided into two sections: One prior to the instauration of "HNC" and one after "before" covers from 1986-1987 to 1900-1991 and after covers from 1991-1992 to 2014-2015. These averages will be obtained for two reasons: The first is to "stabilize" the data, as percentage difference can change a lot from one year to another. The other reason is to compare pollution before and after *HNC". These results will be concrete evidence to see if the law helped decrease pollution levels, because it compares before and after.

Average processed data only focuses on the pollutants, as this can tell us if they have decreased after HNC was created

Table 3: Average Percent Difference of each pollutant before and after the implementation of "Hoy No Circula"

	Pollutant				
	Carbon Monoxide (CO) Levels Concentration (PPM)	Nitrogen Dioxide (NO2) Concentration (PPB)	Ozone Concentration (PPB)	Particulate Matter 10 (PM$_{10}$) Concentration - yg m^3	Sulphur dioxide (SO2) Concentration (PPB)
Before Hoy No Circula (1986-1987 to 1990-1991)	4.52	-1.24	-0.27	-20.9	4.82
After Hoy No Circula (1991-1992 to 2014-2015)	-6.25	-1.67	-1.56	-4.63	-7.69

Graph 1: Percent Difference for S02, PM10, Ozone, NO2, CO before and after the implementation of "Hoy No Circula"- Before: 1986-1991. After: 1991-2015.

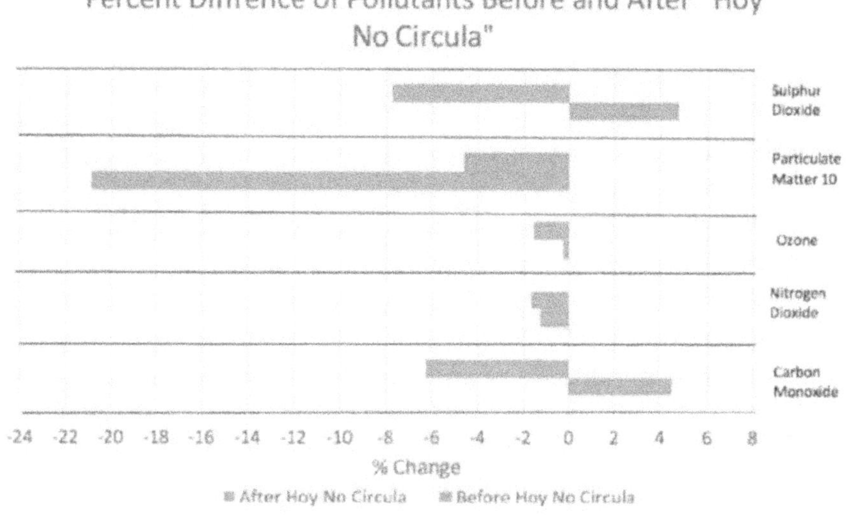

Graph 1 illustrates the average percent change for all five pollutants before and after the implementation of "HNC". It is important to consider that some averages are skewed, as the government began publishing air quality data 4 years before "HNC." This means that the average for the "before" part is not as reliable, as it is only made up of four numbers, and could therefore be subject to other variables not considered in this experiment, such as factories. The area regarding the "after" of the implementation is more reliable because it has averages for over a decade. eliminating the chance of outside factors. One thing that stands out of this data is the fact that, after the implementation of "IINC". all pollutants decreased by at least 1%. Carbon monoxide and Sulphur dioxide were increasing before the program. and decreased after it. Ozone and nitrogen dioxide were already beginning to decrease before the program, but this process sped up afterward. The only outlier is PM. which slowed down after the program. It's an outlier because of lack of data. PM was measured for the first time in 1990, giving the "before" section only one year to create an average. Because it only had one sample, this is not reliable data.

Percent of cars that did not circulate once a week, Justification:

Equation: (Cars that did not drive once per week/ Total Number of cars) x 100

Example: % cars for 2010

Cars that did not drive once a week: 941.687

Total number of cars: 4.166,756

941.687/4.166,756 = 0.22744

0.227744 x 100= 22.8%

Meaning: the 22.8% means that out of all the cars in Mexico City in 2010, 22,8% of these could not drive once a week.

Importance: Illustrates the number of cars affected by the program % change was chosen because it demonstrates the effect of "HNC". It is dependant on the program, not on the number of cars.

Table 3: Number of cars Vs. Number of Cars that did not circulate, percentage of cars that did not circulate per year.

Year	Number of Cars	Number of Cars that do not circulate once a week	Percentage of cars that did not circulate
1986	1,519,411	0	NA
1987	1,549,203	0	NA
1988	1,677,570	0	NA
1989	1,711,583	273,853	13.9%
1990	1,977,554	321,409	16.3%
1991	1,919,557	309,929	16.1%
1992	2,257,243	384,746	17.0%
1993	2,566,043	462,898	18.0%
1994	2,016,038	443,635	22.0%
1995	2,132,325	469,322	22.1%
1996	2,067,206	454,790	22.0%
1997	2,100,283	457,862	21.8%
1998	2,541,957	566,856	22.3%
1999	2,631,169	589,328	22.4%
2000	2,511,543	560,074	22.3%
2001	2,407,362	536,842	22.3%
2002	2,321,702	513,416	22.1%
2003	2,260,123	180,810	8.00%

2004	2,556,032	580,219	22.7%
2005	2,696,220	609,354	22.6%
2006	3,079,690	692,945	22.5%
2007	3,423,719	777,184	22.7%
2008	3,922,587	898,231	22.9%
2009	4,120,535	935,361	22.7%
2010	4,166,756	941,687	22.6%
2011	4,396,912	1,125,610	25.6%
2012	4,615,276	1,186,131	25.7%
2013	4,960,630	1,264,960	25.5%
2014	5,012,750	1,283,265	25.6%
2015	5,259,553	1,341,192	25.5%

Table 3 illustrates how the percentage of cars that did not circulate per year. The older the program, the higher the percentage of cars that did not circulate. This is because the norms became more strict. 2003 is an outlier, as the % of cars not circulating dropped from 22.1% to 8%. This is because the government allowed many more cars to circulate. From 2011 on, the % spiked, increasing from 22% to 25%. (mentioned in context). Because of this, less cars circulated.

Nitrogen Dioxide Vs. Percentage of Cars that did not Circulate

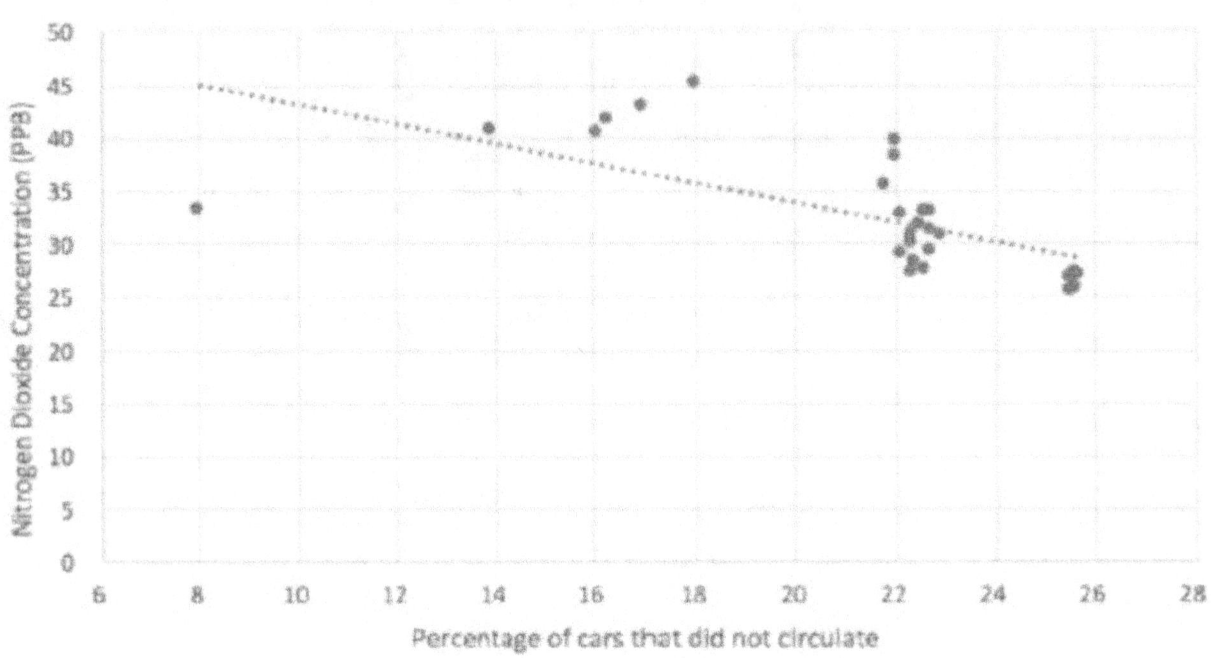

Ozone Concentration Vs. Percentage of Cars that did not Circulate

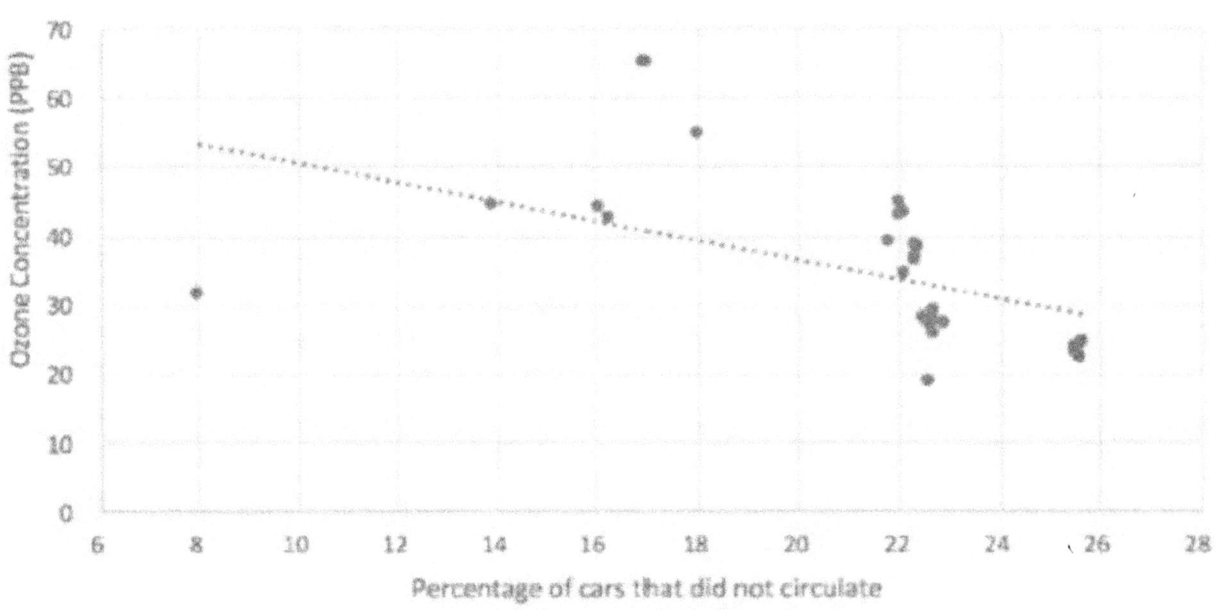

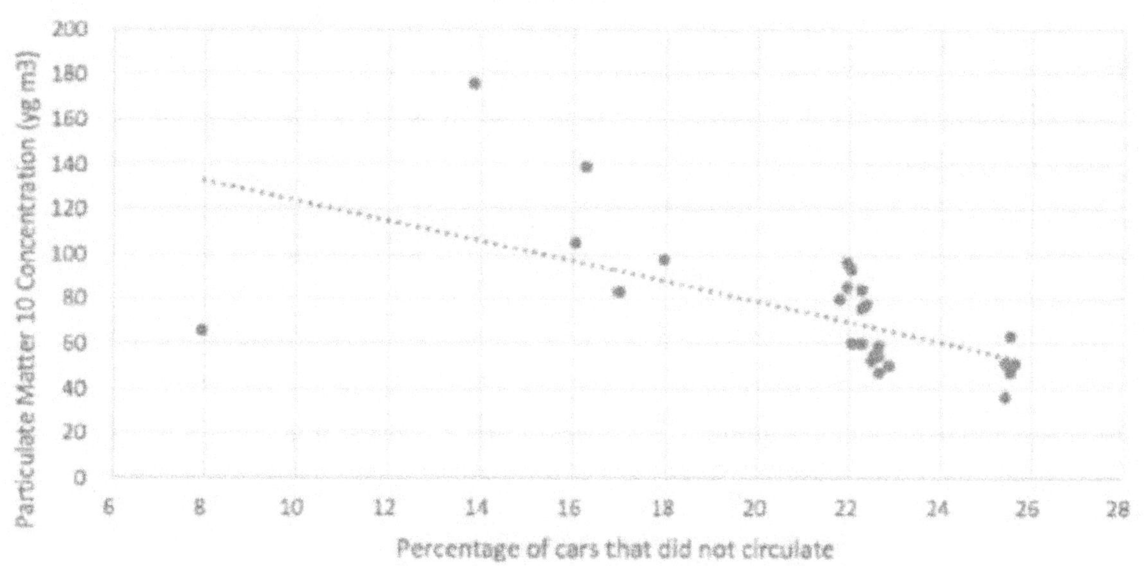

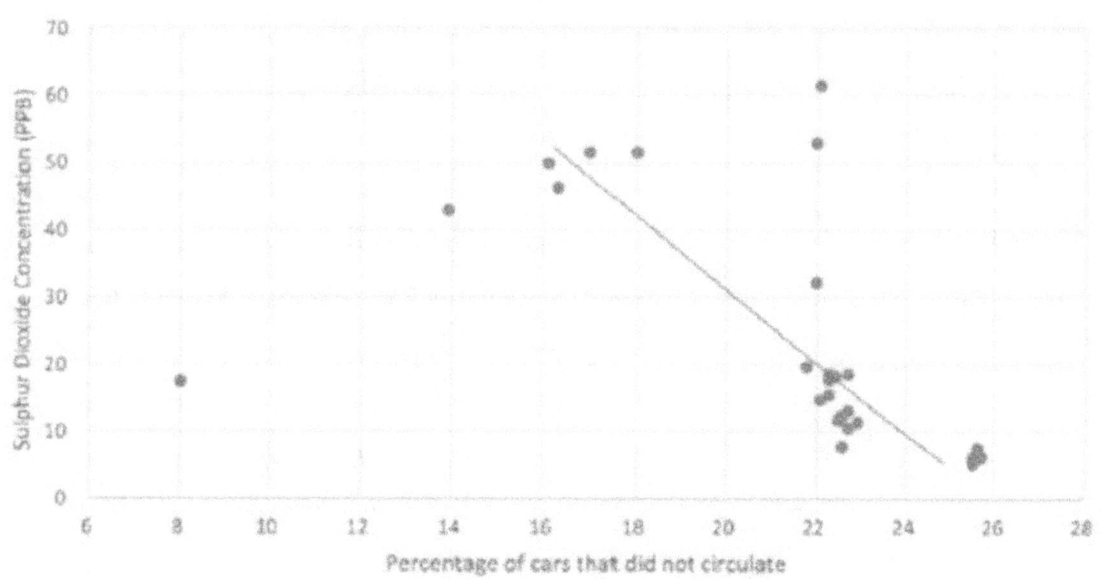

Graph 2: Carbon Monoxide concentration levels (PPM) Vs. Percentage of Cars that did not Circulate once a week for that same year.

Graph 3: Nitrogen Dioxide concentration levels (PPB) Vs. Percentage of Cars that did not Circulate once a week for that same year.

Graph 4: Ozone concentration levels (PPB) Vs. Percentage of Cars that did not Circulate once a week for that same year.

Graph 5: Particulate matter 10 concentration levels (yg-m3) Vs. Percentage of Cars that did not Circulate once a week for that same year.

Graph 6: Sulphur dioxide concentration levels (PPB) Vs. Percentage of Cars that did not Circulate once a week for that same year.

*Outlier- The point in 8% refers to the 2003 change of procedure, mentioned above.

Graphs 2-6 above show a somewhat clear relationship between the number of cars and pollution levels. For all five pollutants, as the percentage of cars that did not circulate increased, the pollution levels decreased sharply. In sulphur dioxide, a change of 2% regarding car circulation equated to a change of 10 ppb. Line of best fit for the other four contaminants are not as steep. meaning that the change in % did not affect pollution change as much. Nonetheless, the data suggests a clear relationship between the two variables. As soon as "HNC" was implemented the percentage of cars that could not circulate increased, and as a consequence, pollution levels drop. It can be concluded that the number of cars, and therefore the implementation of "HNC", has a direct relationship with the amount of pollution. It can be concluded that the less % of cars circulating, the lower the pollution levels.

Discussion and Evaluation:

The results of the investigation illustrate that if "HNC- is implemented. then the concentration levels for CO, S02, PM_{10}, NO2 and Ozone will decrease, supporting the hypothesis. So what does this mean with regards to the environmental issue of air pollution? What impact does that have on society? Graph 1 showed how, after this law was implemented, the concentration of pollutants decreased at a faster rate than before. Graphs 2-6 demonstrated a clear relationship between the % of cars that were kept of the streets to the levels of pollution. All this is important because it shows that the program "HNC" has been effective. This law was created to lower air pollution in Mexico City, and it has been effective in doing so. It could be said that some pollutants have decreased slowly (Ozone decreased by -1.56 since the program began), yet it still shows progress. This is significant because it increases the quality of life and the health of more than 7 million people living in Mexico City. It is impactful because this pioneer program can serve as a model for other cities around the globe. As this lab demonstrates that the program has been successful in lowering the concentration levels of pollutants, it could signal a viable option for cities suffering from pollution problems.

However, one must be careful about drawing conclusions from this lab. It is a small investigation, with weaknesses and limitations:

Error/ Limitation	Impact on the lab	How to fix it
Data Collection (years)	Data collected began in 1986, and the law was created in 1989. This gives very little room to create the average for the "before" percent difference, making this part less accurate, as it could be affected by outside variables,	Look for data from before. This is very hard to do. as the government reports began in 1986. Finding data prior from private investigations is the best option
Pollutant	The 5 pollutants were chosen because these are produced by cars. This does not take into account pollution caused by factories. This impacts the lab because one must also take into account that the decrease in pollutants is not just because of "HNC" -- There are other factors, such as limiting the factories. It makes the data less reliable because other laws apart from "HNC" can affect pollution levels.	Look for pollution levels produced specifically by cars. This is hard to obtain, as the pollutants get mixed up in the atmosphere and one cannot know where they came from. There could be reports concerning pollution from cars specifically.

Applications

This lab focuses on air pollution, which affects cities all over the world. This is partly caused by cars. The government has already taken the first step in mitigating these pollutants, yet the job is still far from done. The most effective way would be to educate the population. By teaching Mexicans the effects that air pollution can have on your health, as well as the benefits of using methods of green transport, the number of cars would decrease. However, this would take a long time, as educating 9 million people can take years. and helping people change their lifestyle could take decades.

Another solution would be to increase public transportation. By building more metro lines, less people will use cars, switching to public transport. This would also take at least a decade to build, and would cost the government millions. Both solutions have limitations, as they would take years and require heavy spending. However, the most effective solution would be to educate the population of the negative effects of riding a car. This education would incentivise people to using other methods of transport, drastically decreasing pollution levels in the long run.

Works Cited

"Dirección De Monitoreo Atmosférico." *Dirección De Monitoreo Atmosférico*. Gobierno De
 La Ciudad De México, n.d. Web. Jan. 2017.
 <http://www.aire.cdmx.gob.mx/default.php?ope=Z6Bhnm1>.

"Calidad Del Aire 2014." *Dirección De Monitoreo Atmosférico*. Gobierno De La Ciudad De
 México, n.d. Web. Jan. 2017.
 <http://www.aire.cdmx.gob.mx/descargas/publicaciones/flippingbook/informe_anual_cali
 dad_aire_2014/#p=28>.

"Mexico City Population 2017." *Mexico City Population 2017 - World Population Review*.
 N.p., n.d. Web. 24 Feb. 2017.
 <http://worldpopulationreview.com/world-cities/mexico-city-population/>.

Riveros Rotgé, Héctor. "Análisis Del Programa "HOY NO CIRCULA"." *Análisis Del
 Programa*(n.d.): n. pag. *UNAM*. UNAM. Web.

"Kognity IA." *The American School Foundation, A.C.* N.p., n.d. Web. Jan. 2017.
 <https://asf.kognity.com/study/app/ess-sl-2017/assessment/internal-assessment-guide/intr
 oduction/>.

Davis, Lucas W. "Driving Restriction and Air Quality in Mexico City." *Driving Restriction
 and Air Quality in Mexico City | Resources for the Future*. N.p., n.d. Web. Jan. 2017.
 <http://www.rff.org/blog/2008/driving-restriction-and-air-quality-mexico-city>.

Molina, Mario J. "Air Quality in the Mexico Megacity:.." *Google Books*. N.p., n.d. Web. 24
 Feb. 2017.
 <https://books.google.com.mx/books?hl=en&id=JLj52dgYKaEC&dq=Air%2BQuality%2
 Bin%2Bthe%2BMexico%2BMegacity%3A%2BAn%2BIntegrated%2BAssessment&print
 sec=frontcover&source=web&ots=_1BYM5I7vY&sig=4Hz_8Ey4bxvAQEIWSchgYZQJ
 0sk&sa=X&oi=book_result&ct=result&redir_esc=y#v=onepage&q=Air%20Quality%20i
 n%20the%20Mexico%20Megacity%3A%20An%20Integrated%20Assessment&f=false>.

Molina, Mario J. "Implications of Air Pollution on Health in Mexico City's Metropolitan Area
 and the Potential Benefits of Applying Control Measures." (n.d.): n. pag. Print.

Edina Rodríguez, Salvador. "El Alarmante Crecimiento De Autos." *Nexos*. N.p., n.d. Web.
 Jan. 2017. <http://labrujula.nexos.com.mx/?p=305>.

Molina, Mario J. "Evaluación Del Programa Hoy No Circula." (n.d.): n. pag. Centro Mario
 Molina. Web.

Cantillo, Víctor, and Juan De Dios Ortúzar. "Restricting the Use of Cars by License Plate
 Numbers: A Misguided Urban Transport Policy." *DYNA*. 2006, Revista DYNA, n.d. Web.
 Jan. 2017.
 <http://www.scielo.org.co/scielo.php?script=sci_arttext&pid=S0012-73532014000600009

"Mexico City Drastically Reduced Air Pollutants since 1990s." *The Washington Post*. WP Company, 01 Apr. 2010. Web. Dec. 2016. <http://www.washingtonpost.com/wp-dyn/content/article/2010/03/31/AR2010033103614 2.html?sid=ST2010033103622>.

'Diagnostico De La Movilidad En La Ciudad De Mexico." *Diagnostico De La Movilidad En La Ciudad De Mexico*. Fideicomiso Para El Mejoramiento De Las Vias De Comunicación Del Distrito Federal, n.d. Web. Feb. 2017. <http://www.fimevic.df.gob.mx/problemas/1diagnostico.htm>.

Corydon Ireland, Harvard Staff Writer |, Sue McGreevey and Mike Morrison, MGH Public Affairs |, Brett Milano Harvard Correspondent |, Christina Pazzanese, Harvard Staff Writer |, Ekaterina Pesheva, Harvard Medical School Communications |, and Alvin Powell, Harvard Staff Writer | "Coming up for Air." *Harvard Gazette*. N.p., n.d. Web. Jan. 2017. <http://news.harvard.edu/gazette/story/2014/10/coming-up-for-air>.

La Visión De La Ciudad De México En Materia De Cambio Climático Al 2025. N.p.: World Resources Institute, n.d. Mexico City Government. Web.

Waller, Ray A., and Francisco Guzmán. "MEXICO CITY AIR QUALITY RESEARCH INITIATIVE: AN OVERVIEW AND SOME STATISTICAL ASPECTS." Los Alamos Laboratory, n.d. Web. <http://permalink.lanl.gov/object/tr?what=info:lanl-repo/lareport/LA-UR-91-2762>.

"Air Quality Surveillance." *Mexico City-Harvard Alliance for Air Quality and Public Health*. N.p., 08 Oct. 2014. Web. Feb. 2017. <https://www.hsph.harvard.edu/cdmx/about-us/air-quality-surveillance>.

"TSP, PM10 and PM10/TSP Ratios in the Mexico City Metropolitan Area: A Temporal and Spatial Approach." *Journal of Exposure Analysis and Environmental Epidemiology*. U.S. National Library of Medicine, n.d. Web. Feb. 2017. <https://www.ncbi.nlm.nih.gov/pubmed/9857289>.

"Vehiculos De Motor Registrados En Circulacion." *Vehculos De Motor Registrados En Circulacin*. INEGI, n.d. Web. 24 Feb. 2017. <http://www.inegi.org.mx/est/contenidos/proyectos/registros/economicas/vehiculos>.

"Mexico - Mexico City Automatic Air Quality Monitoring Network Database." *Mexico - Mexico City Automatic Air Quality Monitoring Network Database | GHDx*. N.p., n.d. Web. 8 Jan. 2017.

<http://ghdx.healthdata.org/record/mexico-mexico-city-automatic-air-quality-monitoring-network-database>.

"The Effect of Driving Restrictions on Air Quality in Mexico City." *The Effect of Driving Restrictions on Air Quality in Mexico City | CEGA | Center for Effective Global Action.* N.p., n.d. Web. Feb. 2017. <http://cega.berkeley.edu/evidence/the-effect-of-driving-restrictions-on-air-quality-in-mexico-city/>.

7. GENDER EFFECT ON WATER-USAGE BEHAVIOR IN TEENAGERS AND ITS EFFECT ON MALLORCA'S WATER SUPPLIES

Author: Anonymous
Moderated Mark: 27/30

Environmental Issue

Water usage and supply strain in Mallorca

Research Question

To what extent does gender play a role in water showering usage hence contribute to water sustainability problems in Mallorca?

Background

In Mallorca, our water supply comes from underwater sources, mainly, this is a non-renewable finite water source and we will eventually use up our resources, if we continue this usage rate. This problem is both local and worldwide, therefore, trying to reduce our usage will help the global community. I have chosen to look at showering as it is the third main water-usage reason, after toilets and clothes washers (3). I chose to look at gender differences to see how both genders differed in their tendencies and compared against each other, as the study conducted by shower power campaign (2) showed that women, on average, were taking 38 seconds longer to shower. Mallorca is in an area which will be seriously affected by global warming (5) hence finding a solution would help my home island. This local issue if not aided will add up to the global issue, and the best way to aid the world water crisis is helping local.

Three main showering factors that increase water problems on the island: length of showering, number of people who take baths and frequency of showering and baths

Based on a guide by "homewaterworks" it claims that the duration of a shower impacts directly the water-usage, obviously, a 10-minute shower will use double the amount of water than a 5-minute shower, hence longer showers will use more water if the flow rate is the same. Also, the number of times showering impacts the amount of water used, as more times showering will always result in more water used. Longer and more frequent showers result on a larger water usage, so a bigger strain is imposed on our water supplies. Also, taking baths will impact our water supplies, as they use an average of 75 litres, obviously, taking baths more frequently will impact negatively as these 75 litres will be used more often. By using more water, we are increasing the water sustainability problems hence straining our water supplies more heavily.

Hypothesis

I believe that based on the information read above, women will be more likely to use more water hence cause a bigger impact on our water supplies. As a result, they will place a bigger strain in our local water supplies and as a result in the world supplies.

Experimental Variable	Named Variable	Equipment or Procedure for Measurement/Control
Independent variable	Gender	Online survey question
Dependant variable	• Shower frequency • Shower length • Bath? • Bath frequency	Online survey with questions related to the variables
Controlled variables	Number of males and females surveyed (92 useful) (100 answers)	Online survey link sent to friends via WhatsApp
	Social/economic class	Financial status varied between all surveyed people, but all were economically stable and able to afford to shower and bath comfortably
	Environment and residency	All surveyed people where residents from Mallorca
	Range of age levels (13-20)	Available to friends

Materials

- Online survey: https://goo.gl/forms/Ts0s0fbihX4NTnuY2 : this allows the survey to be anonymous and quick to answer hence result in more answers

- School and friend social circles: this ensured that the people answering the survey were relevant therefore limiting useless answers

- Internet connection: this made the survey accessible and paperless

- Computer or some form of device able to open the link

Method

1. Safety/ ethical considerations: No safety issues applicable. To avoid bias in answers, the following procedures were followed:

   1. Anonymous: No part of the survey recorded any name or personal information that could link answers to people.

2. Question display: All questions were described in a way that they would avoid bias answering a certain answer due to social expectation

3. Online answering: Due to the survey being online there was no pressure on the answers given by pressure of the investigator as the survey was answered in their intimacy

2. Google forms was used - quick and easy to set up and download and analyse data. The survey had the following six questions collecting open categorical answers, written in both English and Spanish due to my social circles being bilingual.

   1. What's your gender? / ¿Cual es tu género?

   2. How old are you? / ¿Que edad tienes?

   3. How often do you shower? / ¿Cuantas veces te duchas?

   4. On average, what's the duration of your showers? / ¿Cuanto duran tus duchas, de media?

   5. Do you take baths? / ¿Te haces baños?

   6. If you answered yes, how often do you take a bath? / Si has respondido si ¿Cada cuanto te das un baño?

3. The survey was sent out, via WhatsApp messaging application, with a brief rationale and assurance that answers were anonymous and that no personal data would be recorded. Open for two days. Potential 200 answers, expectation was 50% reply.

4. The survey should be open an extra two days if a minimum of 75 answers are not received.

Survey Screenshot:

Water Usage Survey

Trabajo de uso de agua

What's your gender? / ¿Cual es tu género?

○ Male/ Varón

○ Female/ Mujer

How old are you? / ¿Que edad tienes?

○ 13

○ 14

○ 15

○ 16

○ 17

○ 18

○ 19

Data collection and processing

<u>Sampling division and relevant results</u>

The table below shows the results by the surveyed people to all the relevant questions, 100 people answered the survey in total. Anyone who did not answer the question about age was deleted, as it meant that they were either above or below the given ages and as the study was interested in teenager tendencies, they were irrelevant. As a result, there were a total of 94 valid answers, 48 males and 46 females. The interest of this investigation was comparing gender tendencies, therefore, the first and last male were discarded as an equal number of each were used, leaving 46 of each gender. The sampling method used was random as the questionnaire was sent out thus resulting on not knowing who answered due to the questionnaire being anonymous, this method was used to avoid possible bias to answer a question due to social pressure.

Questions		Males: Blue Females: Red		
How often do you shower?		Every other day	Every day	More than once a day
	Number of teenagers	3 4	36 38	7 4
	% (rounded)	7 9	78 82	15 9
On average, what's the duration of your showers?		Less than 5 minutes	Between 5 and 10 minutes	More than 10 minutes
	Number of teenagers	9 6	28 17	9 23
	% (rounded)	19 13	61 37	20 50
Do you take baths?		Yes	No	
	Number of teenagers	12 33	34 13	
	% (rounded)	26 72	74 28	
If you answered yes, how often do you bath?		Once a month	Once a week	More than once a week
	Number of teenagers	6 4	6 16	1 13
	% (rounded)	46 12	46 49	8 39

<u>General trends</u>

- Both, male and female, shower mainly once a day

- The majority shower for at least 5 minutes or more

- Showering once every other day is very uncommon

- Showering for less than 5 minutes is also uncommon

These trends were calculated via percentages, dividing the number of people that answered with the given response by the total number of surveyed people, then, multiplying by 100.

For example:

28 ÷ 46 boys showered between 5 and 10 minutes

28 ÷ 46 = 0.608

0.608 x 100 = 60.8%

Rounded to 61%

The following pie charts display the answers for both, male and female:

Question: How often do you shower? / ¿Cuantas veces te duchas?

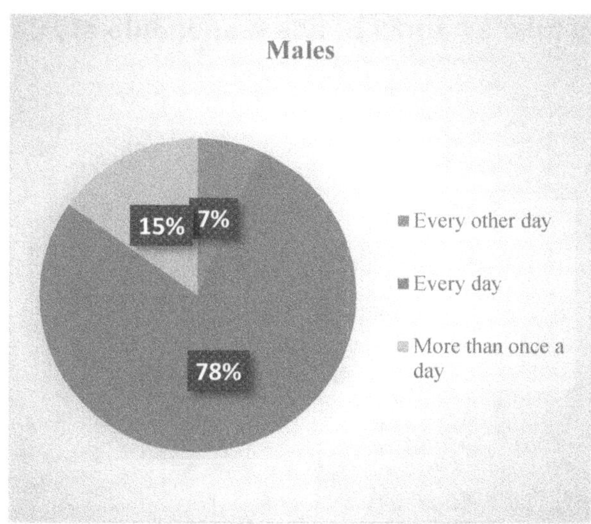

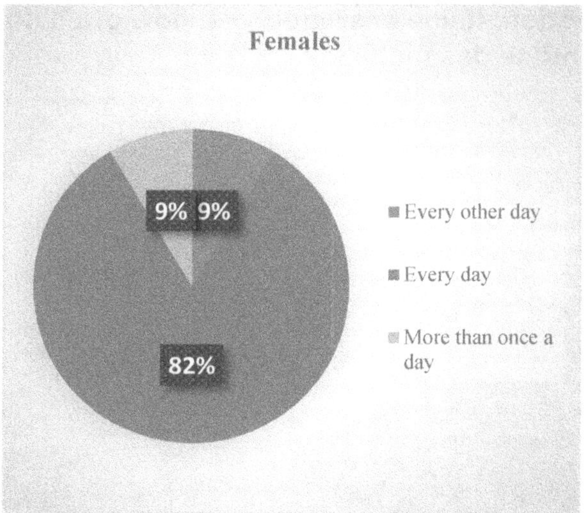

Question: On average, what's the duration of your showers? / ¿Cuanto duran tus duchas, de media?

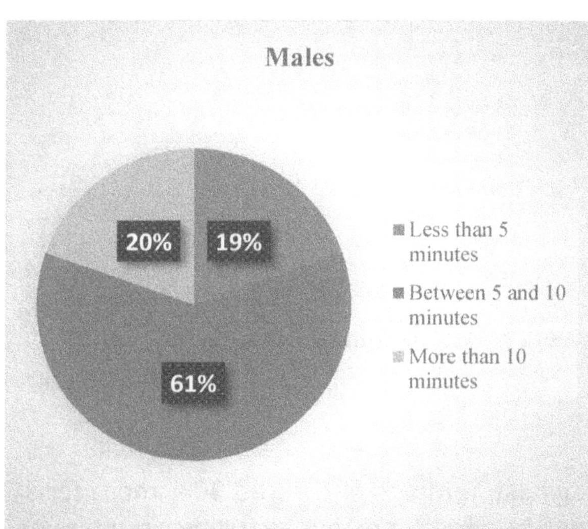

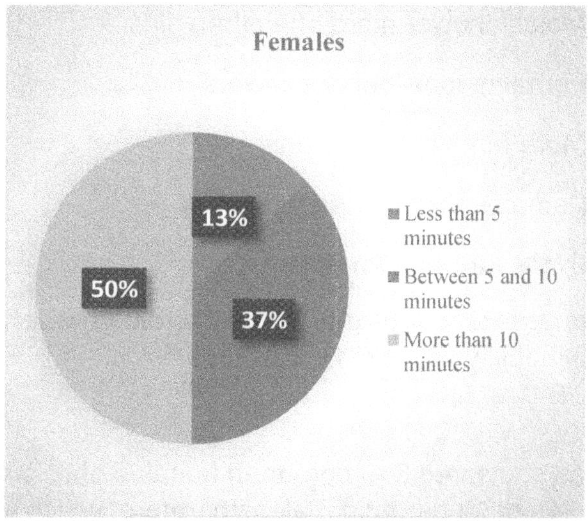

Question: Do you take baths? / ¿Te haces baños?

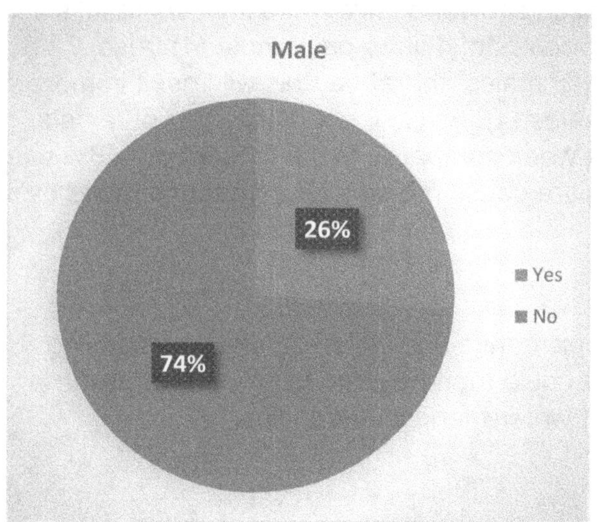

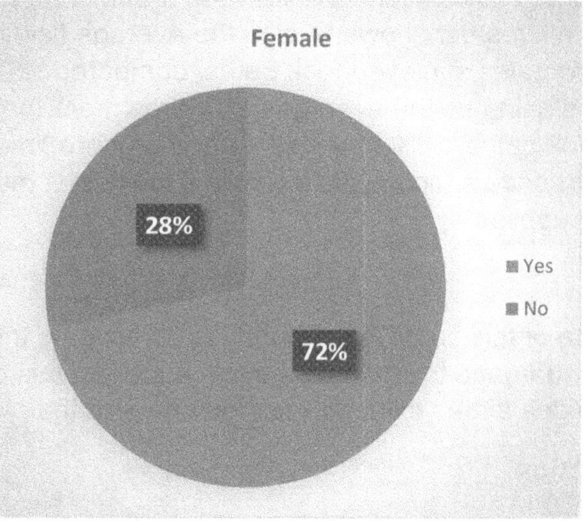

Question: If you answered yes, how often do you take a bath? / Si has respondido si ¿Cada cuanto te das un baño?

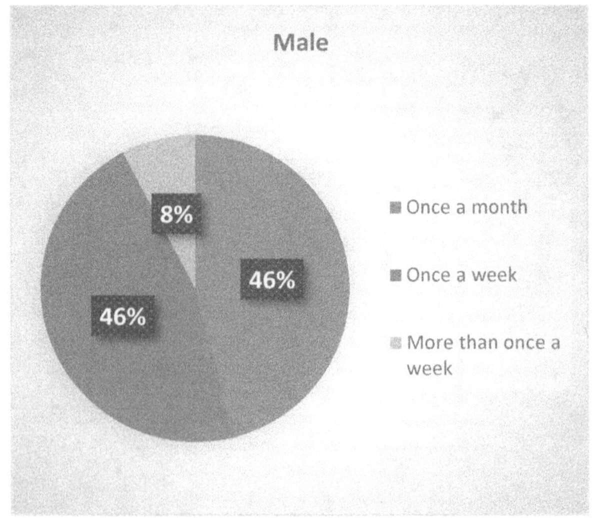

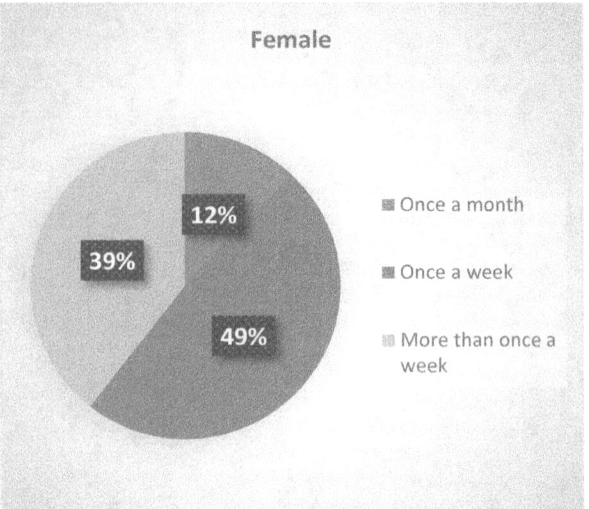

<u>Trends by gender</u>

- Males shower more often than girls

- Females took longer showers

- More females take baths

- Females take baths more often

Analysis and conclusion

To what extent does gender play a role in water showering usage hence contribute to water sustainability problems in Mallorca?

Males showered 6% more than females once a day, whilst, females showered 30% more for 10 minutes or longer than males, therefore, which of these two tendencies is resulting on a bigger strain in Mallorca's water supply? As the difference in the number of times showering is not as significant, then longer showers will result on a bigger water usage – therefore the females will increase the local water usage and place a bigger stress on the water supply. If we do the appropriate calculations, we can see that the difference in showering five more minutes results in 20 more litres per shower, using the average flow of showers of 4 litres per minute (4). Also, 72% of the females surveyed took baths, compared to 27% of males, therefore, this will result in more females using up the average of 75 litres more than males (4). As well, out of the 72% of females who took baths, 88% took baths once or more times a week compared to the 27% whom 46% use the bath once a month. As a result, it looks that males are causing a smaller damage to Mallorca's water supplies.

The use of this survey is to purely to investigate the topic hence find which gender was causing greater damage to Mallorca's water supplies, this data could then be used to prepare awareness campaigns aimed towards each gender and their least water-friendly tendencies.

Evaluation

<u>Errors and limitations</u>

Although as many limitations as possible were taken into account whilst designing the investigation, some are inevitable, or with hindsight, ignored as they were not realised. The first one is a systematic error, the population size, it is a given fact that the bigger the population the more accurate the answers are, as they represent a wider spectrum of answers hence accurate results. To overcome this problem there are various solutions: the first one would be to personally spread the survey more via social media. The other would be to ask surveyed people to share it. Both of these options would solve the problem, as there would be a wider yet valid population. The second error, a random error, was the trustworthiness of answers, due to the fact that the survey was open for two days anyone with the link could have started entering random answers which would distortion the results from the reality hence making the whole investigation useless. The only option to avoid this was to place trust on the people who got the link, although, an extra measure was prepared to avoid this happening, this was the following: when an answer is sent there is a time stamp, therefore if multiple answers were sent in a brief time period, all except the first one were deleted, gladly this strategy did not need to be used. The downside of using this strategy was that maybe multiple people processed their answer in that period and valid data would have been lost.

<u>Factors for further analysis</u>

This experiment took into account showers: their frequency and duration. But it ignored a major point which impacts how much water is used, this being if the showers have any water-saving systems, EG: water-saving shower heads. Therefore, if further, more accurate analysis wanted to be found, this factor should be taken into account as it impacts directly the strain on water supplies. This can be done by adding two extra questions: Do you apply water saving techniques at home? Which techniques do you apply?

As described in the previous section, a wider population would allow more accurate statements and the ability to prove the conclusion further.

Discussion/Solution

Water over-usage is a global problem which is worsening by year. We have already reached a critical state where the government has proposed water restrictions from 2018 onwards (1). This investigation has allowed me to see how gender differs showering/bathing tendencies and how they link to our local crisis. Gender seems to play a role in showering tendencies, and it can be concluded that females cause a bigger strain on the water supplies in Mallorca as they shower for longer and take more baths and more often. I believe that this data should be used to spread awareness about the subject and help people limit their usage for our common-wellbeing. Altering teenagers showering habits should not be too complicated as they are easily influenced by social media and iconic figures. The following proposals would help relief Mallorca's water supplies at my social level, this being school.

- Have an iconic local figure campaign towards a more water-friendly future. EG Rafael Nadal, he is a well-known and respected athlete therefore if he campaigned stating he reduced his water usage many fans would follow his example and try to imitate him and reduce their water usage.

- Segment general gender needs: Shower less often for males and take fewer baths and shorter showers for females, this will result in more effective campaigns with improved results whatsoever method is used.

- Social media campaigning, it is a given fact that teenagers spend a vast amount of time in their social media, therefore targeting them through there with catchy and easy tips would see a massive reduction on the trends proven here. This method would be useful as it would target them directly and result in passive yet useful advertising.

These kinds of campaigns have been heavily used in many advertising campaigns and, as at the end we are advertising a way of living these methods will result on effective reduction of water over-usage in teens and a result relief Mallorca's water supplies.

<u>Works cited</u>

"El Gobierno Se Plantea Restricciones De Agua A Partir Del 2018 Si No Llueve". *El Periódico*, 2018, https://www.elperiodico.com/es/medio-ambiente/20171013/el-gobierno-se-plantea-restricciones-de-agua-a-partir-del-2018-si-no-llueve-6350160. Accessed 16 Nov 2018.

Gillespie, Ed. "Let's Talk Dirty … How Long Do You Spend In The Shower? | Ed Gillespie". *The Guardian*, 2018, https://www.theguardian.com/environment/green-living-blog/2009/sep/04/power-shower-blog. Accessed 16 Nov 2018.

"Showers | Home Water Works". *Home-Water-Works.Org*, 2018, https://www.home-water-works.org/indoor-use/showers. Accessed 16 Nov 2018.

Serle, Jack. "How Long Does A Shower Have To Be, To Use The Same Amount Of Water As A Bath?". *Science Focus - BBC Focus Magazine*, 2018, https://www.sciencefocus.com/science/how-long-does-a-shower-have-to-be-to-use-the-same-amount-of-water-as-a-bath/. Accessed 16 Nov 2018.

"Water Crisis | World Water Council". *Worldwatercouncil.Org*, 2018, http://www.worldwatercouncil.org/en/water-crisis. Accessed 12 Dec 2018.